Why You Love Music

Also by John Powell

How Music Works: The Science and Psychology of Beautiful Sounds, from Beethoven to the Beatles and Beyond

Why You Love Music

From Mozart to Metallica—
The Emotional Power of
Beautiful Sounds

JOHN POWELL

LITTLE, BROWN AND COMPANY
NEW YORK BOSTON LONDON

Little, Brown and Company
Hachette Book Group
1290 Avenue of the Americas, New York, NY 10104
littlebrown.com

First edition: June 2016

Little, Brown and Company is a division of Hachette Book Group, Inc. The Little, Brown name and logo are trademarks of Hachette Book Group, Inc.

Lyrics from "Bad Moon Rising" written by John Cameron Fogerty, reproduced by permission of Concord Music Group, Inc.
Lyrics to Bob Dylan's "Make You Feel My Love" reproduced by kind permission of Bob Dylan Music Co.

Library of Congress Cataloging-in-Publication Data
Names: Powell, John
Title: Why you love music: From Mozart to Metallica — the emotional power of beautiful sounds / John Powell.
Description: First edition. | New York; Boston; London: Little, Brown and Company, 2016. | Includes bibliographical references and index.
Identifiers: LCCN 2015039641 | ISBN 978-0-316-26065-7
Subjects: LCSH: Music appreciation. | Music — Psychological aspects.
Classification: LCC MT90 .P68 2016 | DDC 781.1/1 — dc23 LC record available at http://lccn.loc.gov/2015039641

10 9 8 7 6 5 4 3 2 1

RRD-C

Printed in the United States of America

For Kim

Contents

Why You Love Music

CHAPTER 1

What Is Your Taste in Music?

Your musical taste says quite a lot about you. In the hands of a psychologist, a list of your ten favorite pieces of music can reveal details about how extroverted you are, what sort of background you're from, and even how old you are. The age estimate is pretty straightforward because a surprisingly large proportion of your favorite music will have been produced when you were in your late teens and early twenties. The more psychological aspects of your "musical profile" are the result of decades of research into musical taste and our emotional responses to music. Some of the results of this research have been surprising, to say the least. In this book I'll be looking into the ways in which music affects our lives, from how it brings tears to your eyes, to how it can be used to make you buy more drinks in a restaurant. Let's begin by looking at your taste in music.

At last count there were about 7 billion people on the planet, which of course means that there are 7 billion personalities, all with their own musical likes and dislikes. Faced with the huge numbers involved, the early researchers into people's taste in music decided to make things easier for themselves by reducing the number of personality types in the world to just two: the posh and the rabble. They then went on to work out that there were only two types of music: posh and lowbrow. From there it was but a short step to come up with the startling revelation that posh people liked posh music and the rabble liked all the other stuff.

How refreshingly simple life was in the mid-twentieth century.

Thankfully, things have moved on a bit since then.

Since the 1960s a lot of research has concentrated upon the problem of how to assess people's personalities reliably. Everybody's personality is made up of a combination of several traits, and by the early 1990s psychologists were starting to agree that there were five basic personality traits that can be reliably measured.[1] These are:

Openness (also referred to as Culture or Intellect)
Conscientiousness
Extroversion (or Energy)
Agreeableness
Neuroticism (or Emotional instability)

Recently a sixth trait, a combination of honesty and humility, has been considered for inclusion in the list.

The point of all this is that it's possible to get a rough description of someone's personality by ranking them, on a scale from one to ten, in these five or six characteristics.

In order to match these personality traits to musical tastes, the music psychologists found it useful to divide all the various types of music up into a small number of categories.[2] After a lot of analysis, they found that the musical genres could be grouped together like this:

Reflective and complex music includes classical, jazz, folk, and blues.

Intense and rebellious music includes rock, alternative, and heavy metal.

Upbeat and conventional music includes pop, sound tracks, religious, and country.

Energetic and rhythmic music includes rap, soul, and electronic.

OK, so—who likes what?

It has been found that enthusiasts for *reflective and complex* music tend to score high on openness, are generally poor at sports, good with words, and often politically liberal.

Fans of *intense and rebellious* music also tend to score high on openness and are skilled with words—but they are usually good at sports.

Lovers of *upbeat and conventional* music tend to score high on extroversion, agreeableness, and conscientiousness, are good at sports, and are often politically conservative.

Party animals who are keen on *energetic and rhythmic* music tend to score high on extroversion and agreeableness and are often politically liberal.

These are, of course, just general trends. I'm sure there are a few right-wing professional sportsmen who are jazz enthusiasts, and we all know a disagreeable, introverted pop fan or two. But these trends, identified by psychologists Peter Rentfrow and Sam Gosling,[3] are real. They are, as the specialists say, clear and robust. Although you might have been able to guess some of these results, they are valuable because they are not based on assumptions or guesswork; they are the outcome of statistical analysis of thousands of listeners from America and Europe. There may be similar groupings in other cultures, but we don't know yet. This is a fairly new area of research that has looked only at Western listening patterns so far.

One other piece of information from these studies that is clear and robust is that the four musical genre groupings, which link, say, rock, alternative, and heavy metal together, or combine rap with soul and electronic music, really do work as taste groupings. So if, for example, you like blues, there is a good chance that you'll enjoy the genres that are grouped with it—jazz, folk, and classical.

The fact that we can divide music up into genres tells us that the link between personality and musical taste can't be purely musical. If an entirely music-based choice were involved, then your taste would be for a particular type of sound, but within each musical genre—even within a single album—the range of musical sound is enormous. When I was seventeen, the fourth Led Zeppelin album was considered the pinnacle of heavy rock by all of us who cared about such things (it's the album with "Stairway to Heaven" on it). If we look back at the album from a distance, however, it's obvious that of the eight tracks involved, only four and a half of them are actually heavy rock. The third track, "The Battle of Evermore," is pure folk rock (complete with mandolins), and the first half of the famous "Stairway to Heaven" is an acoustic ballad that has the temerity to involve a couple of tootling recorders! Back in 1971 my friends and I had to go through a lot of angst and recalibration before we could accept the fact that the recorder was now a cool instrument rather than a wooden joke.

Speaking of angstful teenagers, many researchers have noticed that we form strong and loyal links to the music we listen to during our late teens and early twenties.[4] As everyone over the age of thirty knows, your early adulthood music preferences are influenced by more than just the sound of the music involved. A lot of important stuff happens to you during late adolescence and early adulthood: you get your first taste of independence, first kiss, first hangover, first this, first that. It's a time of life when you have to work out your likes and dislikes. Psychologists Morris Holbrook and Robert Schindler have shown that during our late teens/early twenties, we consolidate our preferences for a wide range of things, from genres of novels to types of toothpaste.[5] Naturally, our preferences are not created in a vacuum. Most of us deal with this difficult set of choices by aligning ourselves with a group of friends, and eventually we come up with a workable version of who we want to be.

As a young adult, you generally want to have prestige, or at least warm acceptance in your group of friends, and agreeing with them about things you like (clothes, music, etc.) is one way of achieving this acceptance. Certain artists and bands form the backbone of a musical genre—and you and your friends buy into the whole package rather than deciding that this or that track should be excluded from the genre just because it uses recorders. Another important point is that you and your friends can't consider yourselves to be "cool" unless almost everyone else is "uncool." Once you've decided that this or that band is cool, you want it to be your exclusive choice, so it can come as a terrible blow if your favorite band becomes part of the popular mainstream. A survey of thirteen- and fourteen-year-olds has shown that boys are more susceptible to following their friends' cool trends than girls are, as far as music is concerned, but most of us do it to some extent.[6] As music journalist Carl Wilson puts it:

> As much as we avow otherwise, few of us are truly indifferent to cool, not a little anxious about whether we have enough...and that's not being merely shallow: Being uncool has material consequences. Sexual opportunity, career advancement and respect, even elementary security can ride on it.[7]

The distinction between "cool" and "uncool" music becomes less of an issue as we get older, but is so important to a lot of teenagers that it has been used to manipulate their behavior. A number of town councils use "uncool" music to stop groups of young people from hanging around shopping centers, and anywhere else they don't want them to be. This technique has been known as the "Manilow method" ever since 2006, when Sydney's city council discovered the amazing dispersal effect that *Barry Manilow's Greatest Hits* had on groups of teenagers.

Although most nineteen-year-olds are convinced that the music they and their friends listen to is simply the best music there has ever been, it takes only a few seconds' thought (about ten years later) to realize that this can't be objectively true. So let's look at a few things that *are* true about your taste in music—whatever music it happens to be.

Most of us are pretty unadventurous. One of the first things we do when listening to a new piece of music is classify it.[8] Is it bluegrass banjo? Seventies pop? Classical? Or something else? Then, if the piece fits into one of the musical styles we already enjoy, we give it some attention to decide whether or not we want to add it to the collection of music we like. If a piece doesn't fall into the "Oh, yes, I like this sort of stuff" category, then we don't give it much attention. The result of all this is that we end up gathering an ever-increasing number of similar musical pieces into our "like" categories, and the chances of enjoying a new category of music become pretty small unless we put some effort into it.

As you gather more examples of music you enjoy, you build up a model, or prototype, of a "typical piece I like." The closer any new piece comes to this prototype, the more easily you will accept it. Clearly, you won't be restricted to only one prototype; you'll develop a range of them within your favorite genre—slow/romantic, fast/exciting, etc.—covering various emotional states. And, of course, you can love a number of different genres. Having a wide range like this allows you to choose different types of music depending on what mood you are in and what you are doing.

One of the main things we take into account (consciously or subconsciously) when we are choosing something to listen to is its capacity to stimulate, or arouse, us. ("Arousal" in this context means the opposite of sleepiness.) We tend to be aroused by complex, loud music with a fast beat, and calmed down by simpler, slower, and quieter music. In this context, "complex" doesn't necessarily mean intellectually challenging; it just means that there is a lot going on for your brain to process. For example, fast banjo music is complex, but very few people would say that it was intellectual.

Psychologists Vladimir Konecni and Dianne Sargent-Pollock have looked into our responses to complexity and arousal in music. They found that people preferred to listen to simple music rather than complex, arousing stuff when they were trying to solve problems.[9] Their perfectly reasonable conclusion from this result is that your brain is acting like a computer, and if you run complex programs in the background, you'll find that the main problem-solving program runs more slowly.

You will know from experience that sometimes we deliberately choose music that amplifies the emotional state we are in, and other times we choose music that does the opposite. If, for example, you are excitedly getting ready to go out to a party, you might put on some stimulating music to build the excitement even further. By contrast, if you've just gotten back from said party and are feeling excited because someone you fancy asked for your phone number, you'll choose to listen to something soothing and relaxing to help you calm down.

Our brain doesn't like to be either over- or under-stimulated, and listening to any sort of music involves the brain doing some work. So if you are on a boring motorway drive, you will find that complex, arousing music will reduce the boredom and help to keep you alert. If you then have to drive through a busy city center, you'll probably find yourself turning the music down, or off, or changing to something less demanding, because your brain is having trouble dealing with the complex music and the traffic at the same time.

When we are not dealing with difficult situations like driving in bad traffic, we have a Goldilocks attitude toward the complexity of the music we enjoy. We like it to be not too complicated, not too simple (boring), but just right—and this brings us to an interesting point about music we hear for the first time.

When we hear a piece of music for the first time, it has an extra layer of complexity as far as we are concerned because it's new to us. We don't know what chords or notes are coming next, and our brain is kept busy assessing what's going on. If we like the piece immediately and listen to it repeatedly, it becomes familiar, and its perceived complexity is reduced.[10] Familiarity can cause certain pieces of music to fall below our "too simple" threshold—one of the reasons why some popular radio singles sound great until we've heard them twenty times, and we suddenly lose interest. The reverse can happen with music we initially rate as being too complicated. After a few listenings, the complexity reduces and the piece might then drop into our "just right" category. This has happened to me on several occasions when two tracks—one of which was immediately approachable and the other rather complicated—have appeared on the same album. Initially I have played the album to hear the charming, easy track and I've just had to put up with the more difficult one. Then, as I've become familiar with the album, my preference has changed from one track to the other because the pleasant track has become rather boring and the difficult one has become easier and more rewarding to listen to.

As we get older, although we retain a love for the music of our youth, we often get more sophisticated in our tastes. We start taking more pleasure from things that are interesting rather than those that are simply nice. In music, as in most things, your "too simple" threshold rises as your experience grows. Sophistication doesn't automatically imply poshness, but music that is regarded as posh, such as classical and jazz, often provides a fairly high level of mental stimulation compared to pop music, which is why many music fans become attracted to these genres later in life, even if they weren't keen on them when they were under thirty.

Apart from the complexity and arousal level, another factor that influences what we choose to listen to, or how happy we are with someone else's choice, is how appropriate the music is to the situation in which it's being played. I don't think I'm alone in feeling that the "Bridal March" ("Here Comes the Bride") is one of the dullest pieces of music ever written, but we all smile and nod when it's played at weddings because it has become the standard sound track to the bride walking down the aisle.

Richard Wagner wrote this uninspiring dirge as part of his 1850 opera *Lohengrin,* and it would have sunk into the obscurity it deserves had it not been for the fact that Queen Victoria's daughter, who was imaginatively named... Victoria, chose it as her wedding march (maybe the princess wasn't expecting married life to be much fun). Her choice of music for leaving the church (Mendelssohn's "Wedding March") is considerably more upbeat, but it does sound more like the accompaniment to a military victory than a romantic one. Nevertheless, in those days everything the royals did was instantly fashionable, and so everyone who was anyone started using the same music for their weddings.

That's how it all started, but everything should have changed (at least for the church leaving music) when the Duke of Kent married Katharine Worsley in 1961 to the far superior, joyful sound of the Toccata by Charles-Marie Widor. The royals have been doing their best in this direction ever since: Princess Anne in 1973, and Prince Edward in 1999 both chose the Widor Toccata for their weddings. So why are the rest of us still putting up with tunes that sound as though they could have been written by a six-year-old with a xylophone? Why isn't Widor's Toccata floating out of every church every weekend? Why hasn't a jollier tune replaced the "Bridal March"? I'll tell you why: it's an international conspiracy of lazy organists! The traditional marches are very easy to play, while Widor's Toccata is decidedly tricky. Which is why, when savvy brides ask for it, they get excuses like "We lost the music behind a radiator in nineteen seventy-four" or "I would be delighted to play it but don't you find it a little... vulgar?" So come on, brides—put your foot down and do your bit for music.

Now where were we? . . . Oh yes, taste . . .

Professors Adrian North and David Hargreaves have discovered that we are surprisingly sensitive to the appropriateness of a piece of music to a given situation.[11] In fact, appropriateness is as important as the complexity/arousal aspects I was just discussing in shaping our response to music. While we tend to be forgiving of a dull tune if it fits with our idea of what's suitable (as the "Bridal March" demonstrates), we find inappropriate music highly irritating, as shown by our reaction to badly chosen background music in shops and restaurants.

At this point you might think that I'm going to go off the deep end about background music in the same way that I vented my spleen all over the wedding march. But no, it's more complicated than that.

Let me ask you two questions . . .

1. Do you like background music?
2. Does background music influence your behavior?

Like me, you may well have raised a defiant "No!" in answer to both these questions, and, like me, you are probably wrong. It may be that we are irritated and uninfluenced by *inappropriate* background music—but cleverly chosen background music is generally preferred to silence, and it affects our behavior to a level that is almost laughable.

Professors North and Hargreaves put music speakers on the top shelf of an end-of-aisle wine display in a supermarket to see if different sorts of music could influence the choices we make.[12] The display consisted of four shelves, each of which had a French wine on one side and a German wine on the other. The wines on each shelf were matched for price and sweetness/dryness, so there was a fair competition between the two countries.

Then all they had to do was change the music occasionally and monitor which wines were bought when each type of music was playing.

The results were astonishing.

When they played German music through the speakers, the German wine sold twice as fast as the French stuff.

When they played French music, the French bottles sold *five* times as fast as the German ones.

This implies that we are as helpless as krill in the path of a blue whale as far as marketing music is concerned. And the effect is subconscious: only one in seven of the wine buyers realized that the music had influenced their choice.

In another wine/music investigation, Charles Areni and David Kim looked at how pop music and classical music affected how much money people spent in a wine cellar.[13] This revealed yet another level of gullibility in us poor shoppers. The pop music didn't have any effect on buying patterns, but the classical music obviously made people feel more sophisticated and affluent. They bought the same number of bottles, but they chose the more expensive wine. And I don't mean slightly more expensive. It was over three times the price!

It has even been shown that background music can influence the perceived flavor of wine. In an investigation into this effect, several groups of people were played one of four different types of background music while they were given a free glass of wine. The mood of the music playing in the background was deliberately chosen to be either powerful and heavy, subtle and refined, zingy and refreshing, or mellow and soft. The psychologists running the test didn't, of course, tell anyone that the background music was important; the drinkers just thought they were at a regular wine tasting. The wine tasters were later asked to rate the wine as to how heavy, refined, zingy, or mellow it tasted. The results showed that people tended to match the flavor of the wine with the mood of the music. For example, people listening to powerful, heavy music (*Carmina Burana*) tended to rate the wine they were sipping as powerful and heavy; zingy, refreshing pop music made the wine taste more zingy and refreshing, and so on. In fact the wine was the same in every case (a cabernet sauvignon), and the wine tasters were barely aware of the background music—they were too busy enjoying their free wine.[14]

It seems that music can also influence how much you enjoy the wine. At a wine tasting in London the guests were given wines numbered one to five during the course of the session.[15] They were asked to comment on the wine in each case and also to nominate their favorite. Over the next hour or so, the music changed gradually from mellow classical (Debussy's "Clair de Lune") to the outright drama of "The Ride of the Valkyries." What the tasters didn't know was that the final wine of the session (wine number five) was the same as wine number one. Once again they matched the mood of the music to the taste of the wine. Wine number one was rated as mellow and soft, but the same wine, presented as number five, with the powerful, heavy music, was rated as . . . powerful and heavy. Also, wine number one was no one's favorite, whereas number five was the most popular wine of all.*

It's amazing that our perceptions can be influenced this easily, but plenty of other research has confirmed that background music affects us much more strongly than we would like to think.

Another supermarket study found that slow music made people spend over a third more than fast music did.[16] The reason for this was that slow music made people walk more slowly, giving them more time to browse and buy. The designer of this study, Ronald Milliman, then went on to look at how music affects our behavior in restaurants, and, sure enough, he found the same kind of result.[17]

* It's possible that wine number five did so well because the wine drinkers were just thinking more positively about *everything* after tasting all that wine. Perhaps more tests are required? And if so—can I come along?

With slow music playing, people spent about an hour over their meal, but with fast music, they wolfed their food down in forty-five minutes. The slow music customers also spent about one and a half times as much on drinks during their meal as the fast music diners. And these results seem to be typical of how we respond to the tempo of background music. The original study was carried out in America in the mid-1980s, but when the experiment was repeated by psychologists in Glasgow fifteen years later, they got precisely the same result.[18] Another study showed that slower music even makes us take fewer bites per minute.[19]

But before all you restaurant owners go dashing out to buy copies of *Leonard Cohen's Greatest Hits,* don't forget that, although the slow music customers spent more, they also stayed around, clogging up the restaurant for longer. In a crowded, popular restaurant, you might find it more worthwhile to play fast music so you serve more people per day.

Our pals Professors North and Hargreaves have also shown that the *type* of background music being played influences our behavior.[20] This time they played different sorts of music on different days in a university cafeteria and then asked the customers to rate the feel of the place. The customers had no idea that the type of music playing while they filled in their questionnaires was part of the experiment; they may not even have noticed it consciously, but it did have an effect. According to the customer feedback, easy listening music made the cafeteria feel down-market, pop music made it feel fun and upbeat, and classical music made it feel sophisticated.

The changes in music also altered how much the customers were willing to spend in the place. The prices weren't actually changed, but the diners (who had already bought what they wanted) were given a list of fourteen food and drink items and asked to note down how much they would be prepared to pay for each one. If people filled in these questionnaires when no music was playing, they valued the total list of items at £14.30. If easy-listening music was playing, this rose slightly to £14.51. Pop music pushed this up quite a bit to £16.61, and classical music (as usual) made people come over all posh and sophisticated—and raised the perceived value of the list to £17.23. So the difference between silence and classical music was £2.93, which is about 20 percent.

The overall research result in this area is that, on average, the right sort of background music in a shop or restaurant can increase turnover by about 10 percent.[21] On the flip side, the wrong sort of music (e.g., rap music in a traditional Italian restaurant) can irritate the customers and make the place feel more down-market, or just wrong. One of the worst things a manager can do is allow the staff to bring their own music in. If the staff are of the same age and background as most of the customers, the situation might work out well for all concerned. What usually happens, though, is that the music is inappropriate for the customers, and to add insult to injury, the staff turn the volume up because they are enjoying it, so your favorite wood-paneled restaurant aimed at the over-fifties acquires the ambiance of a bar aimed at the twenty-three-year-old staff. While I do feel sorry for the staff—they must get driven round the bend by "cozy jazz classics for old geezers to eat to"—they'll just have to console themselves with the fact that they are a lot better-looking than the customers, and won't need to be tucked up in bed with a cup of cocoa by 11:30.

One great thing about your musical taste is that it can always be extended to include new genres. Remember the prototypes I mentioned earlier? The more of them you create in your mind, the more types of music you'll enjoy. You'll have to listen to stuff you don't initially like a few times before the new prototype takes root, but I promise it will be worth it, because you'll be increasing the amount of musical pleasure available to you for the rest of your life.

* * *

Before we go much further, perhaps now's the time for a few general comments about this book.

The information I will be presenting to you in the following chapters is based on a vast amount of research carried out by specialists from all over the world. If you see something that you want to look into further, just turn to the references section at the back of the book, which contains details about where to find the original research.

A lot of the information comes from psychological experiments, and although most psychologists may be in broad agreement about this or that point, you can always find a bunch of them who disagree. Rather than presenting all the different opinions and producing a book that is full of "ifs" and "buts," I have tried to stick to the majority view—as presented in such magisterial tomes as *The Psychology of Music,* edited by Diana Deutsch, and the *Handbook of Music and Emotion,* edited by Patrik Juslin and John Sloboda.

At the back of the book I've included some "Fiddly Details" sections. These are short essays on specific subjects for readers who might want a bit more information.

Finally, if you would like more clarification about something I've said, please feel free to email me at howmusicworks@yahoo.co.uk or contact me through my website: howmusicworks.info. On the website you'll also find some videos, including one of me fooling around with a musical beer bottle and an oboe made from a drinking straw.

CHAPTER 2

Lyrics, and Meaning in Music

The power of lyrics

Back in the nineteenth century, British foreign policy seems to have consisted mostly of shooting people in large numbers. By 1812 we were at war with the Russians, the Swedes, and, as usual, the French. (I've no idea what the conflict with Sweden was about—but I bet it had something to do with the IKEA home delivery service.) The Americans, understandably, felt a bit left out and decided to join in the excitement by also declaring war on us.

After a slow start, the Americans really started to get into the swing of things, and by 1814 the British decided to punish them for their lack of ex-colonial gratitude by setting fire to their cities. In August they burned Washington and the following month moved on to attack Baltimore from the sea. Over a twenty-five-hour period the Brits shot about 1,800 cannonballs at Baltimore's Fort McHenry, and by dawn the only light in the city (apart from the dawn) came from the exploding shells—which illuminated the American flag still flying above the fort.

Watching this deplorable display of pyrotechnics was the American lawyer and amateur poet Francis Scott Key. Like all poets he had paper and pencil always at the ready, and he sat down to pen (or pencil) the "Defence of Fort McHenry," which describes the flag fluttering in the smoke and flames:

O say can you see, by the dawn's early light,
What so proudly we hail'd at the twilight's last gleaming,
Whose broad stripes and bright stars through the perilous fight
O'er the ramparts we watched were so gallantly streaming?
And the rocket's red glare, the bombs bursting in air,
Gave proof through the night that our flag was still there,
O say does that star-spangled banner yet wave
O'er the land of the free and the home of the brave?

This—and three more verses—were set to the tune of an old English drinking ditty with the thoroughly unwieldy name of "The Anacreontic Song," written by another bloke with three names—John Stafford Smith.

And that's how the American national anthem—"The Star-Spangled Banner"—was born. I've no doubt that a lot of modern-day Americans wish that lines three and four were a bit easier to remember—but "the land of the free and the home of the brave" is just the sort of thing you want to hear in a national anthem.

"The Anacreontic Song," however, which had the tune first, would cause some consternation if it were sung at, for example, an Olympic Games medal ceremony. The words are addressed to the ancient Greek poet Anacreon, who specialized in drinking songs, and the second verse goes like this:

Voice, fiddle, and flute—no longer be mute,
I'll lend you my name and inspire you to boot.
And, besides, I'll instruct you like me, to entwine
The myrtle of Venus with Bacchus's vine.

Featuring the Roman goddess of sex (Venus) and the god of wine (Bacchus), these lyrics are pretty much the eighteenth-century equivalent of the Ian Dury song "Sex & Drugs & Rock & Roll," which was a hit two hundred years later.

Many Americans feel a surge of emotion when they hear "The Star-Spangled Banner," and I'm in favor of music having emotional effects. But as you can see, the emotional content of this and just about every other song can be changed completely by applying different words to the tune.

Lyrics add a new dimension to a piece of music. The simple addition of a human voice singing "baby baby baby" can infuse the music with sexy enthusiasm or sadness—depending on the way the voice is being used. Emotion can be conveyed simply by vocal inflection, and we can be moved by lyrics sung in languages we don't understand. (I wonder how many Sigur Rós fans speak Icelandic?) There are even examples of emotionally charged songs in made-up languages that no one understands. Enya's beautiful song "Aníron" from the sound track of *Lord of the Rings* is sung in the invented Elvish language of Sindarin.

Of course, most lyrics tell a story as well as relying on emotional vocalization. From the gentle poetry of "Ace of Spades" to the gritty realism of "Puppy Love," we all have our favorites, and of course we all have lyrics we can't stand. For those of you interested in lyrics that are irritating, ludicrous, or simply insane, I'd like to recommend *Dave Barry's Book of Bad Songs*. In this deeply philosophical contribution to human wisdom you'll find more than you might ever have wished to know about songs like "MacArthur Park" (the one about a cake getting rained on) and "Yummy Yummy Yummy" by Ohio Express. There's also a tantalizing mention of a (possibly mythical) country and western song called "The Only Ring You Gave Me Was the One around the Tub."

Sometimes even the finest songwriters have to take liberties with the language in order to make a rhythm or rhyme work. In some cases this just means that the music requires the singer to emphasize the wrong syllable of a word (e.g., apri**cot** at the end of the fourth line of "You're So Vain" by Carly Simon). But it must have been irritating for Neil Diamond when he had to resort to using the nonstandard word "brang" (to rhyme with "sang") in his song "Play Me."

The power of lyrics to alter our response to a tune was demonstrated in an experiment carried out in 1994 by the psychologists Valerie Stratton and Annette Zalanowski, who played the song "Why Was I Born?" (written by Oscar Hammerstein and Jerome Kern in the 1920s) to two groups of listeners.[1] When the tuneful, pleasant music was played by itself, it cheered people up, but when the lyrics (a sad refrain about unrequited love) were included, it had the opposite effect.

Mind you, we don't always listen to lyrics carefully—and even when we do, we often misinterpret them. This is particularly true of young people, who are the target audience of most pop songs. A survey carried out in 1984 offered people four alternative meanings to pop songs.[2] Only one of these interpretations was true and had been verified as such by the songwriter. The songs in question were fairly obvious about the message they were trying to put across. They included, for example, "You Are the Sunshine of My Life" by Stevie Wonder and "Trouble Every Day" by Frank Zappa. The young people (they were all under thirty) who participated in the study got it wrong on average three times out of four—which is the same success rate as picking answers at random without even hearing the songs. Another experiment found that only one third of the people questioned could correctly identify the fact that Olivia Newton-John's "Let's Get Physical" was about sex; an equal number of people thought it was about sports and exercise.[3]

Happily, we get better at identifying what a song is about as we get older—but that doesn't stop a lot of us from falling into the Mondegreen trap.

The word "Mondegreen" comes from the seventeenth-century ballad "The Bonnie Earl o'Moray." Or rather, it doesn't...

The ballad goes like this (and, by the way, "hae" means "have"):

Ye highlands and ye lowlands,
Oh, where hae ye been?
They hae slain the Earl o'Moray,
And laid him on the green.

In November 1954 Sylvia Wright wrote an article for *Harper's* magazine in which she explained that her mother used to read these words to her when she was a child—but although her mother read the correct words, Sylvia had always thought the final two lines were:

They hae slain the Earl o'Moray,
And Lady Mondegreen.

She went on to suggest that, as there was no name for this sort of mishearing of lyrics, they should be called Mondegreens—which I think is fair enough.

Let's take this opportunity to clear up some long-standing Mondegreen-induced misconceptions:

In the 1969 Creedence Clearwater Revival song "Bad Moon Rising," the message is menacing, "There's a bad moon on the rise," rather than helpful: "There's the bathroom on the right."

In the first line of Desmond Dekker's 1968 hit, he's not eulogizing about the fact that he's going to be having "baked beans for breakfast." Apparently the only thing on the menu is bread. And the song is called "The Israelites," not "Me Ears Are Alight."

At the beginning of our favorite song about flying ruminants, Rudolf is being laughed at by *all of* his reindeer colleagues, not by a single, nasty individual called *Olive*.

Meaning and messages in music without lyrics

The fact that lyrics can tell a story has led some people to believe that music by itself can do so as well. Those of you who listen to classical music will probably be familiar with the idea of "program music," which was very popular in the nineteenth century. This type of music is intended to tell a particular tale the composer had in mind when he wrote the piece. One of the most famous of these is Beethoven's *Pastoral* Symphony, the program of which describes a day trudging around the countryside watching peasants dancing and getting caught in a thunderstorm etc.

I was going to use this as a prime example of a program until I found a much funnier one.

In about 1700 the French composer Marin Marais underwent an operation to remove a stone from his bladder and, being a composer, he thought this would make an excellent subject for a piece of program music—and who are we to disagree?

This is what Marais calls a program:

The appearance of the operating table
A shudder on first seeing it
Resolving to get onto it
Climbing onto it
Sitting on it
Grave thoughts
Tying the arms and legs down with silken cords
Making the incision
Introduction of the tweezers
The stone is removed
Nearly losing your voice
The flow of blood

Removal of the silken cords
Off to bed

All this in a piece lasting two and a half minutes, written for the viola da gamba, a cello-like instrument (with optional keyboard accompaniment).

Clearly, Marais couldn't really expect us to work out that a particular musical sound was meant to convey "Removal of the silken cords." We might stand a chance of guessing that a group of notes communicates "Grave thoughts"—but a group of notes suggesting "Grave thoughts" to one person might conjure up images of Etruscan pottery . . . or sunrise over Bolton to someone else.

Going back to the slightly less bonkers program of Beethoven's *Pastoral* Symphony—during the bit where Beethoven is depicting peasants dancing, the bass line is ludicrously simple, just a repeated pattern of three long descending notes. Beethoven knew that the person playing the bass instrument in a village band was usually the least skilled member of the group, restricted to playing extremely simple stuff while his more melodic friends showed off on their fiddles and fifes. So to help us follow the story, Beethoven portrayed his peasant band with an appropriate dance tune and a very simple bass line. But unless you are deeply familiar with the antics of nineteenth-century Austrian village musicians (as Beethoven was), then all this would simply pass you by—as it did me until I started reading up on it. The symphony is full of imitations of the sounds of nature such as birdsong and thunder, but Beethoven himself wasn't comfortable with the idea of trying to use music to tell a story. He described the work as an "expression of feelings" rather than a musical "painting."[4]

For an example of music successfully conveying specific messages, let's return to nineteenth-century America—which, you will recall from earlier in the chapter, was a place of strife and difficult-to-remember poetry.

At the time when Baltimore was being (ineffectually) pounded by the British navy, the armies of both sides of the conflict had been using bugle calls as signals for several years. The soldiers had to know what each call meant, and there was a lot of confusion because various branches of the army used similar calls; for example, the cavalry used a call that meant "Water your horses," which could easily be confused with the infantry call that meant "Make camp." In 1867 Major Truman Seymour—who was utterly fed up with being mobbed by thirsty horses whenever his infantry got their tent pegs out—drew up a definitive list of about forty calls and all the soldiers learned them. This is an example of a working musical language, in which music carries detailed information.

The most famous American bugle call of all, "Taps," was originally used to tell the soldiers that the beer taps were about to be turned off, and would they please all get off to bed because they had a busy day tomorrow. Over the years its role has changed, and it's now the tune the American army plays at dusk and at military funerals—which brings us neatly to a discussion of the Woodstock rock festival in 1969.

Woodstock—one of the first and one of the biggest rock festivals—happened at a time when most of the world thought the Vietnam War had been going on for far too long. The war was, naturally, particularly unpopular with the type of long-haired, pot-smoking layabouts who attend rock festivals.

Enter Jimi Hendrix, guitar hero and antiwar protester. As part of his Woodstock set Jimi plays a feedback-ridden, distorted version of "The Star-Spangled Banner," which goes on for about three and a half minutes. He alternates between a fairly clear rendition of the tune and a virtuoso display of violent noises from his Fender Stratocaster, imitating screams and bombs going off. And just to make his point absolutely clear, toward the end of the solo he plays "Taps," invoking all those military funerals, all those lives lost.[5] A great antiwar protest, which no doubt helped to end the war while simultaneously providing an enormous boost to the sale of Fender Stratocasters.

So here we have a specific, complex message being sent to a crowd of half a million people using music without words. Of course, the message wouldn't have come through to a crowd of people who didn't know the two tunes involved or the political context in which they were played. Music doesn't carry any messages or stories unless you've been taught beforehand that a particular tune or rhythm means this or that. This need for pre-knowledge is a feature of all methods of communication. Sign language, for example, works only if everyone involved in the conversation understands what the signs mean. Except for systems such as bugle calls and rare cases like Jimi's guitar solo, there is no way of extracting a clear message from purely instrumental music because there is no agreed-upon vocabulary. The visual and literary arts are often used to describe objects, people, and actions, but music isn't used in this way because there is no dictionary for translating musical statements into specific meanings.

But if music isn't a method of delivering meaningful statements, what use is it? And why is it so important to us?

It's going to take most of the rest of the book to answer these questions, but we can make a start here by going back to absolute basics.

When I was preparing to write this book, I read an awful lot of definitions of music, and one of my favorites was in one of the old books huddled around the fireplace of a country pub—put there to make the place look cozy and "Olde Worlde" (although I haven't a clue why Olde Worlde props are needed in a pub that was built in 1637). The modestly titled *Universal Knowledge A to Z* defined music as "the sound obtained by combining sequences or groups of notes of different pitch so that they become acceptable or intelligible to the listener."

Intelligible is the important word here. It means that we use our intellect to make sense of what's going on. But if there's no story, what are we trying to understand?

Before we go any further, have a look at this photo:

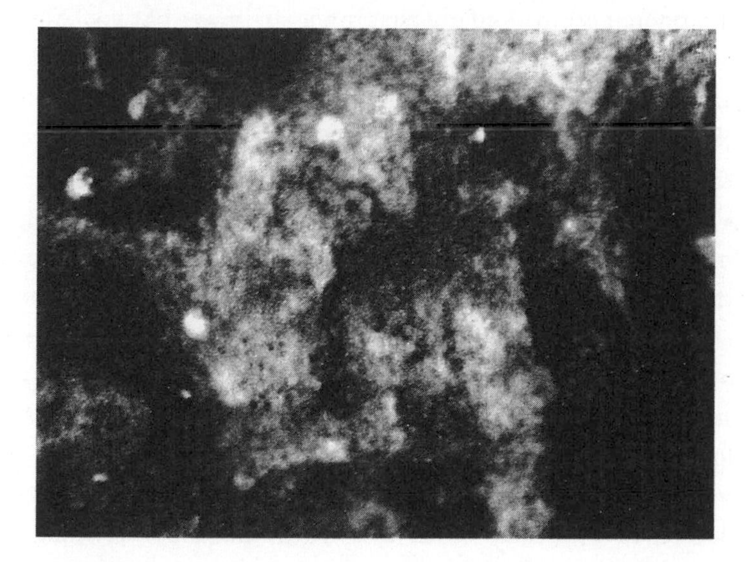

And this one:

Let me guess what just happened in your head...

First of all you wondered what the first photo was about and tried to make sense out of it—and failed. Then you looked at the second photo and worked out that it was an image of a rock formation. Then you may have noticed that some of the rocks look like human faces. Then you looked back to the first photo and tried to make more sense of it—to see if there was a human face in there as well.

Well, I can't guarantee that you saw faces in either photo, but I'll bet you a free romantic weekend for three in Rochdale that you tried to make sense out of the images. You didn't try to interpret them just because they are in this book. You would have tried to make sense of them even if you'd found them behind the sofa, because that's what humans do. We try to make sense out of everything we experience—including music.

When we hear a new piece, we use the large mental library of music we have already heard to give us some context, and we unconsciously construct a set of expectations—we predict that the tune will rise or fall, get louder or softer—and we are often right. We find it pleasant if our expectations are frustrated occasionally, but we don't expect to be wildly donkey. Rather like the way you didn't expect the word "donkey" just then.

Our memories of all the pieces we have heard in the past don't have to be accurate—and we are not looking for perfect copies. It's a bit like watching social interactions and guessing what's going to happen next because you have seen similar situations develop before—so you can see that the girl across the street is about to kiss the man she just met, or that the group of people by the bus stop are listening to a joke and waiting for the punch line.

Most of us are surprisingly expert listeners—and if you need confirmation of this, just consider your ability to tell that a piece of music is about to end, even if you've never heard the piece before. (If you can't tell when it's about to end, it's usually because the composer has put a lot of effort into misleading you. In the late 1970s it was considered very sophisticated to stop a piece suddenly, in the middle of a

. . . but, thankfully, the practice largely died out at about the same time people stopped burning joss sticks at rock concerts.)

Obviously we don't begin life with this ability to compare new experiences to old ones. As babies we have to build up our libraries of "what's going on?" from scratch and we accept new things easily because we have nothing to compare them with. In his book *Sophie's World,* the novelist-philosopher Jostein Gaarder points out that if Granddad starts to hover in the air during a family meal, the baby will calmly accept it while everyone else in the family will be reduced to baffled panic.[6] Babies live in a world of contented, if rather smelly, bewilderment until they gradually build up a set of "this is normal; that is peculiar" rules.

To build up a pattern of expectations, we try to match new sights and sounds to ones we are already familiar with. In the case of the photos, many of us can "see" faces in the second image even though there are no faces there. If you measured up the eye and nose positions of the sandstone "faces," you would find they were nothing like real ones.

And for those of you who are losing your equanimity wondering what the first photo is of, it was just there on my phone one morning. I think it's a flash photo of the fluff inside my trouser pocket. My girlfriend and I spent two or three happy minutes trying to work out what it was, and I thought I'd share the thrill with you.

But let's extract ourselves from my trouser pocket and get back to the music.

Some of your muscles, like your heart and lungs, are designed to keep moving without conscious instruction. Other muscles, such as those in your legs, arms, and hands, generally need to be told to act by your brain. But muscles don't like doing nothing.

Your legs and arms, and various other parts of you, continue to move, even when you're asleep. The reason muscles keep moving is so they can retain some sort of fitness and readiness to obey your next instruction—even if it's only to stretch out an arm to turn the alarm clock off. If your muscles allowed you to switch them off completely, the less frequently used ones would wither away, and eventually you would encounter a situation in which they couldn't save your life.

So there you are: involuntary muscle movement helps you and the rest of the human race to survive.

The same thing can be said of human brain activity. The brain is always desperate for stimulation. Turning off isn't an option. The only way it can retain its ability to keep you alive is by keeping itself in shape by thinking about something or other.

But the brain doesn't like being overstimulated either. Overstimulation leads to panic-like states.

So your brain can't be turned off, and it doesn't like being over- or under-stimulated. As Adrian North and David Hargreaves explain, "the brain works most effectively when moderately aroused; for example, it would be hard to write an essay while feeling sleepy or very anxious."[7]

One of the reasons we love music is that it is an excellent way of providing moderate brain stimulation and giving pleasure at the same time. In Philip Ball's excellent phrase, music "is quite simply a gymnasium for the mind."[8]

But there is, of course, much more to music than its capacity to keep your brain in shape. It can, for example, be a powerful emotional stimulant, as we will see in the next chapter.

CHAPTER 3

Music and Your Emotions

While music (without lyrics) is pretty useless at telling a story, it *can* express and evoke emotions. Before we go any further, I'd like to clarify what we mean when we talk about emotions. Emotions are not the same as moods; we are always in one mood or another, but we are not always experiencing an emotion.[1] Emotions are relatively brief and intense, and they are often linked to unconscious physical reactions such as changes in skin temperature. If an emotional response is music-related, it will be synchronized with the music—that is, the music will trigger the emotion.[2]

Perhaps the most important thing to bear in mind about emotions is that they are biologically evolved reactions which are vital to human survival.[3] From a survival point of view, emotions such as disgust and fear help us to avoid, or run away from, potentially dangerous situations. Anger helps energize us to deal with threats, and happiness steers us toward rewarding situations such as eating and sex. So it's a bit odd that something like music, which has no obvious link to survival, can generate emotional responses—but we'll come to that point later.

Music and emotion are linked in two ways. In some cases we simply recognize which emotion the composer intended us to feel, but we don't engage with it. The other response is when we don't just monitor what's going on; we get emotionally involved. When the "fear" music comes on for one of James Bond's enemies, we don't get fearful; we just think "Ha!—eat metal death, you unpleasant cesspit of moral turpitude!" But if the fear music comes on for James's girlfriend, we get frightened along with her. (This is usually a perfectly justifiable fear, as the average life expectancy of a James Bond girlfriend is about thirty-five minutes.)

The early days of research into the emotional content of music involved a lot of hopeful guesswork as well as some brilliant insights. The pioneering author Deryck Cooke laid a lot of interesting groundwork, but he was over-fond of equating music with language. As we shall see, there *are* some links between speech and music—but they aren't the ones he had in mind. In his book *The Language of Music,* published in 1959, Deryck claimed to have identified sixteen musical devices that had definite meanings. For example, the notes of a major chord, played one after another, from the bottom note upwards, were supposed to convey "an outgoing, active assertion of joy."[4] Unfortunately for the long-term sales of his book, the evidence shows that this type of direct speech–music translation simply doesn't work.

Deryck was particularly keen on the idea that major keys bring happiness and minor keys evoke sadness.* As we'll see in a little while, there is some truth in this idea, but Deryck took a fairly hard line on the issue and therefore found it inconvenient that some cultures—the Spanish and Bulgarians, for example—use minor keys for quite a lot of their happy music. His response to this fact was to argue that the Spanish and Bulgarians were so used to a hard life that they hadn't had time to acquire a "belief in the indi-

* I'll be explaining what major and minor keys are later in this chapter. For this part of the discussion it's only necessary to know that the key of a piece of music is the team of notes being used for that piece. Major keys have a more tightly connected team than minor keys, and major keys are often associated with happiness in Western music. As we shall see, however, this is not always the case.

vidual's right to progress towards individual happiness."[5] In other words, they were so downtrodden and unhappy that they couldn't be expected to create the right sort of (major key) happy music. Even in the 1960s this sort of thinking was easily identifiable as balderdash, and over the next few decades Deryck fell into the obscurity that befalls all those who don't know how to spell their own name properly.

Faced with the complex problem of how music affects human emotions, modern researchers decided to start from scratch—so they began with the fundamental question...

Does music actually generate emotions?

Everyone agrees that music creates emotional responses, but that isn't proof that it's actually true. The psychologists set to work and soon established that people are pretty good at spotting the *intended* mood of a piece of music, even if it's performed "deadpan" by a computer. In one test, researchers asked five composers to write six pieces of music that were individually meant to portray joy, sorrow, excitement, dullness, anger, and peace. The pieces were then played by a particularly dispassionate computer, and the different types of emotion were easily identified by listeners.[6] It has since been confirmed that a wide range of listeners generally agree on whether or not a piece of music is intended to be happy or sad and whether it is supposed to be relaxing or arousing. These are four of the basic emotional states, and they are the easiest for humans to spot and project. You usually can tell how relaxed or aroused and how happy or sad people are by their body language—or, on the phone, by their tone of voice, even if they are speaking in a language you have never heard before. But in music, as with body language, although you can spot the basic intended emotion, it's not possible to interpret detailed emotional information. One person might feel a certain piece of music is happy/playful, and another might hear it as happy/proud.

And being able to identify the intended emotion of a piece of music is only half the story. Can music actually produce an emotional response in the listener? In their quest to answer this question, researchers came up with the cunning ruse of simply asking a lot of people about the effect music had on them. Professor Patrik Juslin and his team asked over seven hundred Swedes to tell them about their most recent emotional musical experience.[7] And the results were very encouraging for the music industry.

Everyone questioned claimed to have experienced emotions while listening to music, and more than eight out of ten of them said their most recent emotional response had been pleasurable—with the five most common emotions being happiness, melancholy, contentment, nostalgia, and arousal. But even the opinions of seven hundred Swedes are not proof of the existence of musical emotions; for proof you need to use things that don't have opinions: you need machines.

Psychologists are never happier than when they can plug hapless victims into machines that go beep. Such devices can measure, among other things, your heart rate, skin temperature, the electrical conductance of your skin, and the amount of muscle activity in your face. When you are emotionally aroused, you use your facial muscles to smile or frown, your heartbeat changes pace, and your skin temperature can fall a little. For example, happy music produces greater skin conductance, a lower finger temperature, and, not surprisingly, more smiling than sad music does.[8] It's very difficult to fake these responses (except the smiling), which is why the police in some countries use the same measurements (via lie-detecting equipment) to detect signs of stress. Lots of experiments have been carried out using this type of equipment, and the results have shown that music can indeed create emotions.[9] Interestingly, if music is calm or happy, we tend to get calmer or happier, but if music is designed to carry a negative emotion like anger, we generally just recognize that it's angry. We don't become angry ourselves, unless the music is playing in a movie, in which case the visual imagery and the music can work together to create a heightened response for both positive and negative emotions.[10]

Brain scanning equipment is the most reassuringly expensive of all the machines that go beep. These devices monitor activity levels in various parts of the brain as we change our thought patterns this way and that. We are not yet at the stage where the medics can look at brain scan images and say, "Oh, look—he's thinking about carrots again," but they can spot whether or not we are emotionally aroused. Ever since such equipment became available, psychologists have been examining the effects of various emotions on activity in different areas of the brain, and the results have been fascinating.

The amygdala, apart from being a very high-scoring word in Scrabble, is one of the most important parts of your brain for processing emotions. It's particularly important in generating fear responses (although that's not all it does), and is sometimes referred to as the "fear center" of the brain. A malfunctioning amygdala can result in mental disorders from depression to pathologic anxiety. In 2001, neuroscientists Anne Blood and Robert Zatorre used brain scanners to monitor blood flow to various areas of the brain while people listened to their favorite music. They found that there was an increase in blood flow to the areas of the brain associated with reward and positive emotions, and a decrease in blood flow to the amygdala.[11] So the pleasure centers were working hard and the fear center was taking the day off. Another study, by Stefan Koelsch and his team, confirmed that listening to joyful dance music increased the blood oxygen levels of various emotion-linked areas of the brain and showed a decrease in oxygen usage in the amygdala. (The amount of blood oxygen in any brain location is an indication of how hard that area is working.)[12] The same project then studied the effects of unpleasant, dissonant music and found that it had the opposite effect of pleasant music: the subjects' blood oxygen levels dropped where they had been high, and rose in the amygdala.* Apart from proving that music actually does generate real emotions, this work indicates that music can be used to manipulate the amygdala in cases where it isn't functioning properly. Music therapy, which we'll be looking at in chapter five, can (and does) use pleasant music to calm down an overactive amygdala to alleviate anxiety and depression.

There have, of course, been lots of other investigations involving various types of brain scanners, and the overall conclusion is that the brain responds to music in a similar way to how it responds to other emotional stimuli.

To summarize all this in very simplistic terms:

Pleasant music stimulates the pleasure centers of the brain and calms down its fear center (the amygdala). Unpleasant music does the reverse.

Which type of music creates which emotion?

The tempo, or speed, of a piece of music is the clearest indicator of its emotional content. If the music is fast, it's unlikely to be sad or tender/romantic. But fast music might be happy or angry, depending on factors such as how warm and consonant the harmonies are. Here is a quick summary of how some of the features of a piece of music are often linked to specific emotions.[13]

---

* The pleasant music was happy classical stuff. To make the unpleasant music, the researchers re-recorded the pleasant music twice, at different pitches. They then played all three versions at the same time. It sounds impressively awful, as you'll hear if you go to: http://www.stefan-koelsch.de/Music_Emotion1.

	Happiness	**Fear**	**Anger**	**Tenderness**	**Sadness**
Tempo (speed)	Fast, steady	Fast, varied	Fast, steady	Slow, steady	Slow, varied
Key type	Major	Minor	Minor	Major	Minor
Average pitch	High	High	High	Low	Low
Pitch variation	High	High	Moderate	Low	Low
Harmony type	Consonant	Dissonant	Dissonant	Consonant	Dissonant
Loudness	Medium to loud, steady	Quiet but variable	Loud, steady	Medium to quiet, steady	Quiet but variable

Certain combinations of musical characteristics generally give rise to particular emotional effects. As you can see from the table, most of the happy, angry, or fear-inducing music you hear is fast, with high-pitched notes and tunes that jump around in pitch. By contrast, tender (romantic) or sad music is generally slow with low-pitched notes and far less movement in pitch.

If you're a composer following these guidelines, you can make a happy tune more tender and romantic by slowing it down and playing it on low-pitched instruments (such as cellos). Or you can make a sad piece of music fearful by playing it on high-pitched instruments (such as violins) and increasing and varying the speed. Film score composers—who need to adjust the theme tune of a movie to fit a wide variety of situations—often use these techniques to manipulate our emotions.

But it's important to remember that there are no firm rules in music; these are just generalizations to which there are many exceptions. The *Jaws* music, for example, is low-pitched and terrifying, and I'm sure you can think of plenty of songs that are high-pitched and sad.

The fact that the speed of the music gives the clearest indicator of its mood is the reason why some Spanish and Bulgarian music is cheerful even though it's in a minor key. The music in question is moderately high-speed dance music, and the high tempo removes any sad influence that the minor key might have.

Obviously there are many more emotions than the five main ones listed in the table, and music can evoke more complicated feelings such as nostalgia, pride, and longing. On top of this, there are so many other musical techniques you can use to influence the emotional content that you can't reliably forecast the emotional effect of a piece of music just by looking at how it's put together.

One of the major variables involved is you, the listener. A wide range of factors can influence your emotional response to music, including your age, personality, musical preferences, familiarity with the music, and the mood you are in. My own response to a piece of music turned from moderate joy to speechless irritation in just twenty seconds a few weeks ago. My girlfriend, Kim, was driving us both back from Southampton to Nottingham the day after a friend's annual barbeque. I was sitting in the passenger seat looking at the sky and enjoying the jolly banjo music on the stereo when suddenly a motorway junction appeared out of nowhere, and we were at the head of the queue at the roundabout. We had to join the motorway immediately—but should we be heading east or west on it? In the time-honored tradition of couples dealing with this sort of emergency, Kim raised the point that, as she was driving and I had the map open on my lap, it seemed logical to her that I might have prepared myself for this decision a little earlier, and spent less time staring gormlessly out of the window.

At least that was the gist of what she said.

I stared at the map. I stared at the road signs. I stared at the map again—and found that I just couldn't think clearly with that dreadful jangly bloody banjo music playing. I turned it off and my head cleared immediately. "West," I said—and we headed off into the sunset. Within a few minutes we were back in the correct frame of mind for jolly jangling and turned the music back on. If there's one thing life has taught me it's that banjos and map reading don't mix. This, of course, fits in well with the research by Vladimir Konecni and Dianne Sargent-Pollock I mentioned in chapter one, which found that we don't like complex music when we are trying to solve a problem.

Everyday musical experiences

In the Western world, about one third of the things we do every day are accompanied by music, and for around half of that time the music has some sort of emotional effect.[14] In 2008 our friend Professor Juslin and his team carried out a detailed study on the effect of music on our everyday lives.[15] They asked thirty-two Swedish students to carry around handheld computers that were programmed to beep seven times a day at random intervals. Whenever they heard the beep, the students had to answer a list of questions about what they were doing at that point, how they felt emotionally, and whether or not they could hear music. If they could hear music, there were further questions about the music itself and what effect they thought it might be having on their emotional state.

As I'm sure you'll be delighted to hear, they found that, whether music was involved or not, calm contentment and happiness were the most commonly felt emotions, while negative feelings like guilt and disgust were quite rare. Some listeners experienced music-induced emotions far more often than others, but the addition of music to any situation tended to boost the number of happy or elated experiences and minimize the incidence of anger or boredom.

On top of this, whenever there was a mood change caused by

the addition of music to a situation, it was almost always a change for the better, with the person becoming happier and more relaxed. This is partly, of course, because most music is designed to be pleasant and relaxing.

Now, if the subjects of this study had been from a population predisposed to giddiness—the sort of "carry me to the next bar, when does the carnival start?" party animals you might find in Brazil or Wales—we might be suspicious of the results. But this was a group of Swedes—the inventors of the ultra-safe Volvo, the vegetarian meatball, that most unhappy of all fictional detectives, Wallander,* and his even more miserable father. If music is capable of cheering up the compatriots of Wallander's dad, we can be sure we're on to a good thing.

So now it's official: music *is* good for you. It generally cheers you up and/or relaxes you.

Which brings us to the question of why we sometimes choose to listen to sad music. We've all had the experience of selecting a downbeat album to match our mood after a bad day. At times like this the misery of life on earth feels a little more acceptable with an appropriate sound track: it's you and Morrissey against the rest of the world. But there are other times when we choose to listen to sad music even when we are in a good mood. This sounds like a bad move, but here is what psychologist William Forde Thompson has to say on the matter:

> Sad music, like Shakespeare's tragedies, may be appreciated for its artful construction and for its ability to illuminate an emotion that is an essential part of being human. We can examine and appreciate the sadness conveyed in music while being reassured that there are no actual consequences to face.[16]

---

* The stories are great, though. Have a look at the TV series with Kenneth Branagh—but don't expect it to be a cheery evening's viewing.

So there is a certain amount of comfortable satisfaction involved in listening to sad music when you're not feeling sad yourself—as any country and western fan will tell you.

Background music and your emotions

As we saw in chapter one, much of the music you hear during the course of an average day isn't chosen by you. It's chosen by a committee of salespeople who have decided that this or that selection will optimize your willingness to buy shoes—or persuade you to linger longer over lunch in their trendy new restaurant.

A surprisingly large majority of people are quite happy to be fed a steady diet of music chosen by others, but there is a small minority who really don't like it. These naysayers are concentrated in one social group—males over the age of forty, or, to use their more technical appellation, grumpy middle-aged men. The psychologist's best guess as to why we grumpy middle-aged men don't like background music is that we are used to having control over things around us.[17] We don't like it when we can't choose, so we get tetchy and disagreeable, and we don't like shoe shops anyway, so the irritating music gives us a good excuse to stalk off to the nearest pub.

But about three quarters of us find background music pleasant most of the time, and it generally supports, or moves us toward, a positive emotional state.[18] The music involved is usually straightforward, and its emotional content can be identified very quickly. In some cases it takes only a second or so.[19] Music that is produced specifically as the background to shopping or eating (as opposed to hits of the seventies) is deliberately designed to be non-distracting and utterly unmemorable—so it creates a sort of aural wallpaper, allowing you to focus on the pros and cons of the turquoise suede sandals as compared to the blue kitten-heeled winkle pickers. This need to be unobtrusive is the reason why aural wallpaper music is often made up of old, familiar songs (like "The Girl from Ipanema") played without (attention-grabbing) lyrics on mellow-sounding instruments.

The result of listening to background music is generally a diluted version of your emotional response to listening to music you might have chosen. This could mean a minor reduction in boredom when you're doing something tedious, or a gradual shift from fed-up to not-quite-so-fed-up. But as music psychologist John Sloboda points out, these small changes shouldn't be dismissed as trivial. The accumulation of little positive pushes has been shown to enrich your life by improving how you perform both socially and intellectually.[20]

Useful music

The enjoyment of music seems to be restricted to humans. We have incontrovertible scientific evidence that marmosets don't give a damn about music, and tamarins also have a tin ear.[21] Of all the creatures tested so far, only one has been found to be capable of spontaneous bodily movement in time to the beat of music. Yes, you guessed it—put your hands together and let's hear a big round of applause for the only other creature capable of getting down on the disco floor... the parrot. It's possible that being able to move to a beat is somehow linked to the other capacity that parrots and humans share—the ability to learn how to pronounce words—but the link hasn't been proved yet.[22] The most famous of all dancing parrots can be found on YouTube if you look for "Snowball—our dancing cockatoo." Watching Snowball pick up the beat to Queen's "Another One Bites the Dust" is really impressive—and very funny. Neuroscientist Aniruddh Patel (author of *Music, Language and the Brain*) was astounded when he saw the video, and became even more impressed once he'd carried out tests which proved that Snowball really was responding to the beat.

Part of our enjoyment of music comes from the fact that we can use it to enhance our performance in certain types of activity. Music can be particularly useful in five roles in our day-to-day lives.

Apart from the obvious music-movement link in dancing, we occasionally use music *to energize us and focus our attention* on a physical task. In olden times sailors used sea chanteys to encourage and coordinate their team efforts—pulling on various ropes, splicing mainbraces, whittling wooden legs, and so on. Nowadays the crew are more likely to be using driving, rhythmic music to optimize their activities in the ship's gym. Studies have shown that energizing music in the gym motivates people to stay on equipment longer and to increase their pace to match the pulse of the music.[23] This can be of particular use to long-distance runners, which is why Rule 144.3b of the USA Track and Field organization bans headphones from competitions if awards or prizes are involved. In the less cutthroat world of charity marathons, headphones are often banned but for a different reason—namely the safety risks of not being able to hear what's going on around you, particularly if the roads are not closed. As you'll see if you visit runners' websites, these rules are extremely unpopular with the competitors, many of whom feel that they can't finish a marathon or half marathon without their favorite running music. I do feel sorry for the runners involved but, frankly, if I want to travel thirteen miles I take a taxi.

We can also use music *to distract us* from the tedium of a boring activity, and this is particularly useful if you are doing something dull but not quite mindless. If your job is simply to transfer hundreds of books from boxes to shelves, it may not be terribly boring because you don't need to think at all, and you can daydream or chat while you are doing it. If, however, you have to put the books on the shelves in alphabetical order, then the boredom is monumental: you can't do the task on autopilot, but you are using your brain for only about twenty seconds out of every minute. In this type of situation music is a godsend. Your attention can shift from the music to the task and back again every few seconds, so you feel less bored. To quote a marvelous phrase from psychologist John Sloboda, using music as a distraction "is a way of engaging unallocated attention and reducing boredom."[24] (I love the idea that you can have bits of spare attention hanging around on the street corners of your mind, waiting to be allocated.)

Occasionally we use music *to enhance the meaning of a situation*. Obvious examples of this are weddings and funerals—although it also applies to situations like choosing romantic music for a drive in the country with a new beloved.

And speaking of romance, music has been shown to aid the release of the hormone oxytocin into our blood. This hormone, which is also released during breast-feeding and sex, encourages *bonding* between people.[25] More generally, music encourages social bonding whenever we dance or sing in groups—as any soccer fan will tell you.

Another common use for music is *stress reduction*. We all have our favorite stress reduction pieces, and they don't necessarily involve calm, relaxing music—just stuff we love. Stress reduction has become such a popular branch of music that you can now choose from a broad range of albums specifically designed to relax you. My dentist has one that sounds like a mash-up of whale song, harp strumming, and orchestrated hits of the eighties. I can't even begin to tell you how annoyed and stressed it makes me feel. I must explain to him one day that my flinching and whimpering have nothing to do with the drilling or injections—it's the thought of another twenty minutes of Lionel Richie and Culture Club played on panpipes. Surprisingly, this relationship I have with my dentist's sound system brings us to an important point about emotional states generated by music: in most cases a strong emotional impact can be achieved only after you, the listener, have decided that you are going to encourage it to happen. The music can't do the job on its own. You have to be a consenting partner.

Your earliest musical moment

I can almost guarantee that you won't be able to remember the first time music had an emotional effect on you. Why am I so confident? Well, because it probably happened before you were born. Human fetuses are responsive to sounds almost two months before they are born: fast, rousing music speeds up their heart rate and slow, soothing music calms them down. A few weeks later, as they are getting ready to join the hustle and bustle of daily life, they show emotional responses to music that can be measured by both heart rate changes and differences in how much they move around in the womb.

Once the baby is born, the parents (particularly the mother) will often talk to their babies with different types of sing-songy lilt depending on whether they want to entertain them or calm them down. The soothing phrases—"There, there, sweetie, it's OK"—usually involve low notes, descending melodies, and a very slow pace, rather like actual lullabies. Playful chitchat—"You lovely boy! Where *did* that noise come from?"—uses higher notes, a wider range of notes, strong rhythms, and lots of repetition. These motherly musical offerings, together with actual play songs and lullabies, have an effect on the levels of cortisol in the baby's bloodstream,[26] and your bloodstream cortisol level is associated with how aroused or relaxed you are.

Lullabies are so important to babies that they can even be used in a medical context. In 2000 Dr. Jayne Standley used lullabies to help premature babies who had difficulty feeding. Babies born before the thirty-fourth week of pregnancy haven't developed the coordination to suck liquids without inhaling the milk, so they have to be fed by tube. Babies who can be taught to suck put on weight faster than tube-fed babies. With this in mind, Dr. Standley invented the Pacifier Activated Lullaby system, or PAL.[27] When the baby sucks on the pacifier (or dummy, as they are called in the UK), the PAL plays a lullaby, and if the baby stops sucking, the music stops. The babies clearly enjoy the music, because after only a few minutes of practice they learn how to keep the music going by continuing to suck.

As babies approach their first birthday, they start to develop the ability to distinguish between happy and sad music, and by age four, kids can point to a picture of a facial expression that matches the mood of the music they are listening to.[28] From that age on they are also capable of singing the same song in a sad or happy way—if bribed with enough chocolate. But at this stage in their lives, "happy" simply means quick and loud, whereas "sad" means quiet and slow. (As adults we pick up several more sophisticated clues about the mood of the music, but we still retain this basic link between happiness, speed, and loudness.)

Between the ages of five and twelve we start to accumulate a library of expectations about music, and we can be amused or annoyed by stuff that violates those expectations. For example, once we've been exposed to a lot of tunes and songs, we acquire a subconscious ability to understand what key the music is in. We probably don't even know what a key is, but we can hum the group of notes being used to produce the music—and could, for example, make a good guess at the note the tune will end on and what chords are likely to be used in the harmonies. If unusual or out-of-key notes are used, even kids as young as nine will find it laughable or weird.[29] During these primary school years we also start to identify minor keys with sad emotions and major keys with happiness[30]—which, by happy coincidence, is the next subject we are going to look at.

Why are major keys happy and minor keys sad?

As I said earlier, you are an expert listener to music even if you have never been taught anything about it. Whenever you listen to a piece of music, you are subconsciously analyzing all sorts of things, like what genre it is, what rhythms are involved, and, as I just mentioned, what key it's in—even if you don't know what a key is.

So—what is a key, and how on earth do we identify them if we don't even know what they are?

In Western music we have twelve different notes we can choose from, but we tend to use a selection of about seven of them at any one time (although the composer might change from one group of seven to another during the course of the piece). In any Western music the particular group of about seven notes involved at any one time is called the key.

As the music plays, you build up a clear picture of which seven notes are being used, so that if someone plays a note that isn't in the same key it will sound wrong to you, out of place—out of key.

As you subconsciously gather your collection of seven notes, you will also identify the most important member of the team. This is the keynote. The keynote of C major is C, and its team is C, D, E, F, G, A, and B. The keynote of D major is D, and the full team is D, E, F sharp, G, A, B, and C sharp. Don't worry; you don't have to remember these rows of note names. I'm just putting them in here to show that the groups of notes—the keys—have a fixed identity.

In Western music we use two types of keys, minor and major, and our subconscious music analysis equipment doesn't find it particularly difficult to work out which type we are listening to. As I said at the end of the last section, we associate minor keys with sadness and major keys with happiness—but why?

One of the main reasons is cultural. In northern Europe and the United States, most of the sad lyrics you hear are set in minor keys (e.g., the jazz song "Cry Me a River") and happy lyrics tend to be presented in major keys (e.g., the Beatles' "Here Comes the Sun"). So people who grow up in these societies expect these links. But as usual with music, it's not a hard-and-fast rule even in these societies. It's possible to write happy music in minor keys (like Purcell's Round O in D Minor*) and sad music in major keys (like Leonard Cohen's "Hallelujah").

And as we saw earlier, some societies (Spain, the Balkan countries, and India, for example) use minor keys for joyful *and* sad music. So is there any objective reason why minor keys should sound sad?

Well, there are a couple of technical reasons why minor keys might be considered more suitable than major ones for expressing complex emotions like sadness and longing—but these shouldn't be thought of as rigid laws; they are more like gentle persuaders.

The first reason why minor keys express sadder emotions so well is that they really are more complicated than major keys. To understand this idea, let's take a quick look at how the strings on a harp are tuned to produce a major key.

A harp string produces a musical note by vibrating backwards and forwards at a certain *frequency*. A string might, for example, be traveling from left to right and back again 110 times every second, causing the pressure of the air near our eardrums to go up and down 110 times a second. This fluctuating pressure pushes our eardrums in and out at the same rate—so we hear a note of that frequency. (This is the frequency of the note A_2—the note produced by the A string of a guitar.)

---

* In classical music terms, this piece is a "rondo" or "rondeau" (i.e., it has a main opening tune that it keeps returning to), but Purcell jokingly referred to this particular piece as a Round O. Benjamin Britten used the tune as the basis for his "Young Person's Guide to the Orchestra."

If we make the other strings on the harp vibrate at frequencies that have a simple mathematical relationship to that first note, they will sound good together.* For example, a string vibrating 165 times a second will work well because 165 is one and a half times 110.

For a major scale, you start with a note of a certain frequency and then add notes whose frequencies equal the first note's frequency multiplied by 1⅛, 1¼, 1⅓, 1½, 1⅔, and 1⅞. These simple relationships mean we have put together a strong, closely related "team" called a major key.

Let's stick with the sports analogy for a minute. A minor key is basically a major key with three team members replaced by rookie players who don't quite fit in. Our "1⅔" member, for example, is replaced by the relatively unfamiliar fraction "1⅗." To complicate matters further, minor key music allows some of the original major team members to substitute occasionally for these new, weaker players.

All in all our new team has less cohesion and is therefore more conducive to complex or somber musical moods than the more self-confident major team.

Music psychologists David Huron and Matthew Davies have unearthed another potential reason for our association between minor keys and melancholy. Their vast survey of minor and major key music revealed that, on average, minor key melodies involve smaller jumps between notes than major key tunes.[31] When we are happy we tend to use relatively large jumps in pitch when we talk. The phrase "Hi, glad you could make it! How're you doing?" might usually involve a fairly big jump down in pitch between "Hi" and "glad" and then a jump back up for "How're you doing?" When we are unhappy, we use smaller jumps. The sentence "I'm sorry to hear about your mother. How're you doing?" would

* This principle will be discussed in more detail in chapters nine and twelve.

involve only small jumps in pitch. We can often judge a person's mood from the size of their pitch jumps, even when we can't understand the language they're speaking. So it seems reasonable that the smaller pitch jumps of minor key music could be one reason why we interpret it as sad.

Links between music and speech

A growing amount of evidence shows that there are similarities between the ways we perceive emotion in both speech and music.[32] Even as young children we become very skilled at quickly picking up clues from the speed, loudness, pitch range, and timbre* of sounds, and brain scans have shown that we use many of the same areas of the brain when we are judging the emotions being expressed by music and speech.

Language could only have been developed long after we evolved the vocal equipment required for speech, so there must have been a period of time when early humans used their voices, without words, to communicate emotions (anger, affection, etc.) and basic information ("Something dangerous is coming!" "There is food over here!" etc.). The linguist Dwight Bolinger has suggested that high and rising pitches would have been used to indicate interest ("Look at this!") and incompleteness ("What's going on?"), and, as languages were eventually developed, this resulted in an almost universal use of rising pitches for questions in modern human languages.[33] Similarly, the falling vocal pitches that we find at the end

---

* The *timbre* of an instrument is the quality of the sound it makes. Low notes on a clarinet have a rich, warm timbre. A flute has a clear, pure timbre. The timbre of your voice can vary from warm and relaxed to anxious and stressed. Timbre is more fully described in section A under "Fiddly Details" at the back of this book.

of statements in nearly every language indicate a reduction in interest and a feeling of finality.[34]

You might have thought that early humans developed speech first and then singing, but actually the two have probably always been intertwined. It is not too fanciful to imagine that early humans hummed and lah'd "songs" to themselves and to their babies long before languages were developed. It seems likely that these "songs" would naturally follow the characteristics of the wordless communication techniques outlined by Bolinger, with pitches that rise toward the middle of each (wordless) phrase, then fall toward the end—a pattern that is still followed by many songs to this day.

Most musical phrases follow an arch-shaped contour: they start on a certain note, then rise in pitch before descending again, often to the note they started on. There are two common types of arch, both of which are also used in speech patterns. When David Huron analyzed about six thousand European folk songs, he found a straightforward "up toward the middle—down toward the end" arch in about half of them.[35] And if you pay attention to the way you use your voice, you'll notice that this gradual up-and-down curve is common to many of the statements you make. If you say, "I thought we'd go out for a meal tonight, but I don't know where," you'll probably find your pitch rising up to the word "meal" and then falling away again. Another type of arch sometimes used in music and speech involves a steep jump upwards at the beginning of the phrase, followed by a series of downwards steps. This sort of contour is common in eastern European or Arabic laments and is similar to wails we produce when we are extremely distressed.[36]

Folk songs, which are often centuries old, have given us another clue about the link between music and speech. In normal speech we raise and lower the pitch of our voice from one syllable to the next, but the range of pitches we use is obviously a lot narrower than the one we use for singing. You might therefore assume that there is no direct relationship between the size of the pitch jumps in melodies and those used in speech—but this isn't strictly true, as neuroscientist Aniruddh Patel discovered.

Dr. Patel carried out an in-depth comparison of French and English folk songs to see if the language had an effect on the melodies. He identified rhythmic similarities between the songs and the language in each case—which is hardly surprising since the melody must fit the words, and you would therefore expect the rhythms of the language to influence the rhythm of the music. When he analyzed how the pitch went up and down between words in the two languages, he discovered that there were smaller jumps in pitch in spoken French than there are in English. This tied in with the fact that the French folk songs jumped around in pitch less than the English ones. So the language involving smaller pitch steps (French) also developed songs with smaller pitch steps.[37]

As we've seen, music and speech share a common set of emotional cues: big jumps in pitch, a fairly high degree of loudness, a fast tempo, and a pleasant timbre indicate happiness,[38] and more muted sounds, smaller pitch jumps, and a slow tempo indicate sadness. This is why, as I said earlier, we can often understand the emotional content of a conversation even if we can't understand the words. We are very good at picking up emotional cues simply from how the voice is being used—and it probably all starts with those lullabies and play songs I mentioned earlier.

If you hear a mother and her six-month-old-baby "talking" to each other, it sounds as if the mum is imitating the sounds of the baby, but in fact it's the other way round. The mother (and other adults) will start talking to the baby in what psychologists call "motherese" long before the infant is developed enough to respond with "ma-ma ga-ga."[39] Motherese is different from normal speech in several ways. It is much more repetitive, has a more regular rhythm, is slower and higher pitched, and involves a lot of deliberate emphasis. Also, as music psychologist Diana Deutsch has noted, it often uses the up-down arch common to a lot of musical phrases (e.g., "Go-o-od girl!").[40] All of these features make motherese much more melodic than normal speech. When babies reach an age of about six months, they start trying to copy motherese by babbling. Babbling is very useful because it allows babies to practice using the seventy or so muscles they will need to control in order to demand expensive birthday presents later in life.

The sing-songy nature of babbling and motherese is common to a wide range of human societies, and it's therefore not surprising that lullabies are similar all over the world. Nor is it surprising that, having had these enjoyable musical experiences as babies, most of us grow up with a natural affinity for music.

How does your brain turn music into emotions?

I said earlier that emotions are linked to survival—so why do we have emotional responses to something as apparently remote from survival as music? Well, your brain doesn't treat music as a special case; it just treats it as audible information, a sequence of sounds to be rapidly processed like any other. Part of this processing often involves the generation of an emotional response. Music psychologist Patrik Juslin and his colleagues have suggested seven basic psychological mechanisms by which music produces emotions.[41] They are:

Brain stem reflexes

When you hear a sudden noise—whether it's the sound of a tree falling toward you or an unexpected saxophone wail in a jazz standard—a very basic part of your brain called the brain stem recognizes that something urgent is going on. Before you can say, "What the hell was that?" your brain stem has given you an adrenaline boost and you're ready to deal with the situation: you are in an emotionally aroused state. Then other parts of your brain put a damper on things, effectively saying, "Calm down, you drama queen, it's just music," and you relax a bit. But you won't completely calm down because you need time to recover from the original brain stem response. Also, luckily for you, your brain stem can't be trained to respond more calmly next time, so the stimulation effect works again and again. This is the probable cause of our emotionally excited response to music that involves sudden, loud, and dissonant sounds, and fast or rapidly changing rhythms.

Rhythmic entrainment

Your pulse rate and breathing rate slow down if you are calm and speed up if you are emotionally excited. Under the right conditions your heart rate can rise or fall toward the beat of the music you are listening to, and this fools your brain into experiencing the emotion that is appropriate to your new heart rate. Fast dance music in a nightclub excites and energizes you for a couple of hours, slow blues in a dimly lit bar relaxes and calms you.

The same rhythmic entrainment effect can happen with your breathing rate—but because each breath takes a few seconds, your breathing synchronizes not with the beat of the song but with aspects of the music that last for several beats, such as complete musical phrases. The process of synchronization takes several minutes, and although your pulse or breathing rate might move toward the beat of the music, some beats are too slow or fast for you to match them.

Evaluative conditioning

If a piece of music is repeatedly linked to a pleasant activity such as watching your favorite TV show, then that music will make you happy even if it's not accompanied by the show. You've conditioned (brainwashed) yourself into becoming cheerful whenever you hear that particular piece of music. This kind of conditioning works for other emotions as well. For example, even today, whenever I hear the music for the soccer program *Match of the Day,* I am filled with an almost uncontrollable urge to strangle my brother

Richard. This is because as children, in those far distant days when each family had only one TV set, he and I had to take turns at ten o'clock on Saturday nights, alternating between our favorite shows. As a result I was prevented from watching the ravishing Diana Rigg in *The Avengers* every second week—and if you've seen Diana Rigg dressed in black leather, you'll understand what a torment it was for a red-blooded Englishman in his early teens to have to watch a soccer match instead.

Emotional contagion

Emotional contagion involves identifying what emotion the music is trying to portray and then allowing yourself to be caught up in that emotion. You become infected by the emotion in the same way that you might become cheered up just by being near a bunch of cheerful people. As I mentioned earlier, emotional contagion happens far more often for cheerful or relaxing music than it does for unhappy or angry music.

Visual imagery

Though not everyone has this experience, music is pretty good at encouraging you to daydream in pictures. Many people conjure up images of landscapes and other visual images when they listen to music, which enhance their connection to the music and generate deeper emotional responses. In the 1960s, music therapist Helen Bonny developed a music therapy technique known as Guided Imagery and Music (GIM), in which patients are encouraged to experience (and talk about) visual images while music is playing. This kind of guided musical visualization can be very effective at reducing stress levels and alleviating depression.

Episodic memory

This is the "Darling, they're playing our tune" effect. A piece of music triggers a specific memory, and you then return to the emotional state you were in when the memory was created—the joy of your wedding day, for example. Besides regenerating the original emotional state, this type of stimulus can also arouse emotions such as nostalgia and, in the most acute cases, attacks of revolting sentimentality.

Musical expectancy

As an experienced listener you build up expectations of what the music is likely to do next. Even two or three notes of a piece you've never heard before will prompt a subconscious expectation of what notes might come next. Whether your expectations are confirmed, delayed, or violated, you'll have an emotional reaction, ranging from the smug glow of satisfaction when you get it right, to the thrills you experience if the composer or songwriter delivers something unexpected and beautiful.

So what has all this to do with survival?

Music is not necessary for survival, but emotional responses are, and music creates emotional responses. Let's have a look at a few examples of how the seven psychological mechanisms just listed are useful in keeping you alive in non-musical contexts:

Brain stem reflexes make you jump out of the way when you need to.

Rhythmic entrainment helps you to do repetitive physical tasks more efficiently (no music is necessary—you can choose your own rhythm).

Evaluative conditioning helps you learn what's good for you.

Emotional contagion can be thought of as a version of empathy: it helps you bond with people around you (e.g., mother and child).

Visual imagery allows us to "dry run" things we are thinking of doing to imagine how safe or unsafe they are: "If I climb that tree to steal some bird's eggs, will that branch hold my weight?"

Episodic memory helps you figure out what sort of situations you find rewarding, threatening, tiring, etc.

Expectancy helps you to estimate what's about to happen (in language and in life) and prepare accordingly. This is a crucial skill. Even casual chitchat involves you getting your response ready while your co-chatterer is still finishing her sentence; if we couldn't do this, every conversation would be full of those long, pregnant pauses you get in brainy French films. Those of us who don't live in French films have to forecast how the other person will end each comment, so we can chip in immediately and demonstrate how attentive and clever we are.

All of these mechanisms help you to live. They happen whether you want them to or not, and, by happy coincidence, they can be triggered by music. The mechanisms are independent of one another, so one or more of them can be activated by a certain piece of music, and each one leads toward different emotions, which explains why different people have different emotional responses to the same song. A dance tune might stimulate rhythmic entrainment and emotionally arouse one person, but trigger an episodic memory in others, leading them to feel nostalgic for their school days. The fact that several mechanisms can be activated at the same time could explain the mixed emotions we sometimes experience when we listen to music. For example, if we combine the emotions I just mentioned, we could end up with aroused, nostalgic dancing—whatever that looks like.

This research by Juslin and Co. is only a few years old and may have holes picked in it in the future, but for the moment it's one of the few theories that tries to answer the question of how music generates emotions.

As you can see, the study of music and its effect on our emotions has come a long way since Deryck Cooke's assertion that minor keys are sad, unless you're Bulgarian . . . or Spanish.

CHAPTER 4

Repetition, Surprises, and Goose Bumps

Repetition in music

I hope you enjoy this joke—even though you may have heard it before:

> Question: What do you call a fly without wings?
> Answer: A walk

I hope you enjoy this joke—even though you may have heard it before:

> Question: What do you call a fly without wings?
> Answer: A walk

The previous eight lines tell us something very important about repetition. The fact is that there is no such thing as true repetition as far as a human being is concerned. When you read the joke for the first time, it created a particular response, which might in this case have been a smile of amusement or recognition. When you read it a second time, you may have thought "This must be a printing error" or "Is this any different from the first version?" but you certainly didn't experience a repeat of your earlier smile/recognition response. The print on the page is an exact copy of the earlier rendition, but your attitude toward the lines has changed—simply because they were repeated.

In most contexts (reading, conversation, jokes, etc.) we don't usually value or enjoy repetition. But there is one exception. We love repetition in music, which is why nearly all music is extremely repetitive.

Pop songs are repetitive, classical music is repetitive, rap is repetitive—I could go on...

Of course some "repeats" in music are not really identical. There may be a change of harmony under a repeated melody, or the same melody might be played by a different instrument. But even if the repeat is an electronically produced exact copy, our response to it is adjusted by the fact that we heard it earlier.

This is one of the reasons why musical repeats are not boring. If they were boring we'd have a problem, because there is an extraordinary amount of repetition in almost every genre of music.

You'll have noticed that I said "almost" every genre. The exceptions are rare and usually involve intellectual modern classical composers deliberately trying to avoid repetition at all costs. To investigate our addiction to musical repetition, music psychologist Elizabeth Margulis played recorded excerpts from little-known pieces of this sort of non-repeating music to a roomful of professional music theorists (people who could be expected to be well disposed to new musical ideas such as the avoidance of repetition). Dr. Margulis played some of the excerpts in their original "no repetition" versions, but without telling the audience, she also tampered with some of the pieces, adding repeats of short sections of the music. For example, if the composer had written a piece made up of three segments, intended to be played in the order 1, 2, 3, the audience heard the music with the first and third segments repeated, like this: 1, 1, 2, 3, 3.

The audience members were then asked how much they enjoyed

the various pieces of music, and the results showed that they much preferred the pieces that included the repeats.

As Dr. Margulis says in her fascinating book *On Repeat:*

> This is a stunning finding, particularly as the original versions were crafted by internationally renowned composers and the (preferred) repeated versions were created by brute stimulus manipulation without regard to artistic quality.[1]

So not only do we love repetition in music, but also we get less enjoyment out of music that doesn't involve repetition.

One reason why musical repetition is enjoyable and useful to a listener is that music cannot be summarized. If I say to you, "Steve just rang me to say his brother borrowed his motorbike without asking so he's getting the bus. He says he'll be here about an hour late. He sounded furious," you know that the gist of this message is "Steve will be an hour late." If you are a particularly supportive friend, you might extend the summary to "Steve will be an hour late and he'll need cheering up." But you'll probably discard the information about Steve's transport arrangements. You can't do this summary thing with music. A tune has no gist. If you want to hear, or even remember, some part of a melody, you have to experience it at the speed at which it is normally played. Also, although you can start and finish your memory (or listening experience) wherever you choose in a piece, you can't take shortcuts from one part of a tune to another without ruining the flow of the music.

Music flows through your brain in a stream you can't easily think about while it's happening. So it's very useful if that great melody or bass line you've just heard is repeated a few times. After several repeats you start to understand how the music goes and begin to follow what's going on. As psychologist Peter Kivy puts it, musical repeats "allow us to grope so that we can grasp."[2]

Repetition allows us to concentrate on different aspects of the music and enjoy the experience more deeply. In the case of a song, we might relax our concentration away from the melody (which is repeating) and start listening to the words. In an instrumental piece, we might listen more carefully to the rhythm of the bass notes or the soulful timbre of the sax solo. Elizabeth Margulis suggests that repetition gives us pleasure as the result of "a growing sense of inhabiting the music."[3] There is a parallel here with how small children experience stories. Little kids like stories to be repeated—and they like the repetition to be as exact as possible. If you leave out a detail or give the rabbit the wrong sort of voice, you'll find yourself being treated with withering contempt by a three-year-old. One reason children like the repetition is that they are processing a lot of information, and repetition allows them enough mental space to do so with pleasure. Almost any story for small children will include some words they haven't used before. Psychologists Jessica Horst, Kelly Parsons, and Natasha Bryan have discovered that if children hear a new word as part of a repeated story, they start to use it much faster than if the same word is presented in lots of different tales.[4] This is not an obvious result. You might have thought that hearing a new word in several different contexts would be the best way for a child to get a sense of when to use it. Instead it seems that simple repetition in a single context gives the child a solid base from which to begin using the new word.

Repetition and music are so interwoven that in some cases you can turn a non-musical sound into a musical one simply by repeating it a few times. For a great example of this, check out Diana Deutsch's speech-to-song illusion (you can hear it on her website,* or on YouTube if you search for "speech to song illusion"). Ironically, Professor Deutsch discovered the illusion by accident while

---

* http://deutsch.ucsd.edu/psychology/pages.php?i=212.

she was editing the spoken part of her CD *Musical Illusions and Paradoxes*. At the time of the discovery she was editing out a hesitation in her delivery of this sentence:

> The sounds, as they appear to you, are not only different from those that are really present, but they sometimes behave so strangely as to seem quite impossible.

In order to fine-tune the timing of the sentence, Diana was listening to the phrase "sometimes behave so strangely" on a repeating loop when suddenly the words started to behave strangely...

Although the phrase was spoken normally, after a few repeats it seemed to turn into a little song with a clear melody.

The recording on the website (and on YouTube) consists of the whole sentence followed by repeats of the "sometimes behave so strangely" part. If you listen to it you'll hear the full sentence and the first couple of repetitions of the shorter phrase as normal speech. But after a few repeats, for most people the phrase will start to sound like singing. The really weird thing is that if you go back to the beginning of the recording and listen to the full sentence again, it all sounds like perfectly normal speech except for "sometimes behave so strangely," which sounds like part of a song, at least to the vast majority of us (about 85 percent of listeners). So some normal speech can be turned into song simply as a result of repetition. Professor Deutsch was so impressed with the strength of the effect that she put it on her next CD of aural conundrums, *Phantom Words and Other Curiosities*. (Those of you who are interested in how fallible our hearing systems are will find these CDs mind-boggling.)

Further investigation of the phenomenon confirmed that the pitch jumps of the spoken version were a long way off the pitch jumps of the song it apparently produces.[5] The spoken version does go up and down in pitch, so it has a contour like a tune, but the pitch jumps are not as big as those in the tune that appears after repetition.

During Diana's investigation, people were asked to imitate the phrase after they had heard it only once, when they were still experiencing it as normal speech. When they did so, they followed the pitch jumps of the normal speech version fairly accurately—as you'd expect. Later, after they'd heard several repeats of the spoken phrase and it had turned into a song in their heads, they were again asked to imitate it. This time nearly all of them followed the same contour of up-and-down jumps in pitch, but they increased the sizes of the jumps until they matched real musical intervals: in other words, they were singing. Not only were they singing, but they all sang the same tune. Their brains had taken the real information (the moderate ups and downs of speech) and changed it, magnifying the jumps until they matched musical intervals. And all of this happened simply because the phrase had been repeated.

A much simpler example of repetition transforming sounds into music happens every day, whenever we hear a car's turn signal, or a clock. Most turn indicators and clocks make the same ticking noise all the time—they go tick, tick, tick, tick. But for reasons that no one has figured out yet, we experience this ticking as a repeating cycle of tick tock, tick tock. Try listening to a clock or watch tick. You will probably hear this tick tock, tick tock pattern, with the *tock* seeming a little lower in pitch, or quieter, or louder than the *tick*. Now concentrate and try to group the sounds in threes rather than twos—tick tock tick, tick tock tick. Say this pattern of ticks and tocks aloud (in time with clock) to get yourself into the swing of it, then stop saying it and you will probably continue to hear the new pattern. If you manage this, you will notice that you've turned some of your tocks into ticks, and indeed some of your ticks have transmogrified into tocks. This proves that the ticking sounds must all be identical, and you can now, if you like, make yourself hear what's actually happening—tick tick tick tick tick. But don't get too proud of your newfound skill. This clarity will quickly collapse and you'll soon be on the slippery slope to ticktockitude again. It seems that our brain finds it easier to deal with similar repeating sounds if it divides them into little groups.

A team of French psychologists working at the University of Burgundy has confirmed this. They wired people up to EEG machines and observed their brain activity as they listened to a series of identical tones. They found that the brain automatically organized the tones into pairs and experienced more activity on the first tone in each pair.[6] The sounds were identical, but listeners were subconsciously accenting them into the equivalent of TICK tock TICK tock.

A large number of psychological tests have shown that we experience our most intense emotional responses from familiar music, and repetition is of course a great aid to familiarity, whether it's repetition of the whole piece, the chorus of a song, or even a single note—as in the beginning of "Flash" by Queen.

Even if you don't realize that something in the music is being repeated, it still casts its "repeat" spell over you. The music theorists I mentioned earlier may not have realized that the repeats were increasing their pleasure, and most of us don't listen to music with some sort of conscious repeat barometer, so we don't generally notice all the repetition going on in the music we listen to. In some cases only the harmony is repeating, or the bass line—but whatever it is, it helps to draw us into the music.

The subconscious aspect of all this is very interesting. Back in the 1960s the psychologist Robert Zajonc identified a phenomenon he called the "mere exposure effect."[7] Zajonc's investigations showed that we have more positive reactions to things we have seen or heard before, *even if we don't realize that we have experienced them previously*. This phenomenon sounds similar to the old adage "I like what I know and I know what I like," but it's deeper and weirder than that.

In one experiment, Zajonc showed non-Chinese students twelve meaningless squiggles like these, which look like Chinese characters:

The students were divided into several groups, and although each group was shown the same twelve images, they weren't shown them the same number of times. During the course of the test some images were shown once, some twice, some five times, some ten times, and some appeared twenty-five times. But the images were shuffled around so an image that had been seen ten times by one group was seen five times by the next and twenty-five times by another, and so on. The images were shown to the students for only about two seconds each time, so they had very little chance of remembering any of them.

Later, the students were told that the images were Chinese ideographs for various adjectives, some pleasant (perhaps "beautiful" or "sweet"), and some unpleasant (perhaps "rotten" or "difficult"). They were then asked to rate them just by looking at them—to assign each one a "goodness" score, from zero (if they thought the ideograph might mean something really bad) up to a maximum of six (if they thought it might mean something very pleasant). The students probably thought they were rating the images on the basis of their appearance ("this one looks spiky and aggressive so I'll give it a 1," etc.), but what they actually did was rate the ideographs they had seen more often as the most pleasant.

Although the students hadn't seen the images often enough or for long enough to remember them, in each group the images that had been shown most frequently were rated as having the most positive meaning, and vice versa. Each group of students had different "most positive" and "most negative" symbols because the frequency of appearance of the different images had been changed from group to group. The overall effect was quite clear and has been reproduced many times since in lots of different contexts. Generally, for images and sounds, we respond more positively to what we have been exposed to before and more negatively to the unfamiliar.

The really surprising thing about the mere-exposure effect is that it doesn't seem to have anything to do with conscious thinking or choice-making. It's an unconscious process. The effect has even been shown to work on images that are flashed for only four one thousandths of a second, too fast for the brain to consciously "see" the image at all.

The unconscious mere-exposure effect is probably one of the reasons we enjoy repetition in music, even when we aren't conscious of the repetition. The effect also works to some extent when we hear variations on a theme, as opposed to verbatim repeats. This makes sense, given that a lot of repetition in music involves slight changes of rhythm, pitch, or instrumentation.

So repetition gives us a baseline of pleasure. But for the truly beautiful musical moments, we need the repetition to be spiced with surprise.

Surprises in music

We don't usually pay attention to everything that's going on around us. Attention requires energy, and as biological systems, we like to minimize the energy we use. On the other hand, we need to pay attention whenever something important or interesting is going on. In bygone days "something important" might have been the sudden appearance of a wolf pack, or a thunderstorm. Nowadays it's more likely to be the sudden approach of dessert, or the realization in a boring meeting that someone just mentioned your name. When this sort of thing happens, both your arousal and attention systems kick in together. The arousal system gets your body prepared to deal with the situation by increasing your pulse, glucose uptake, and breathing rates (among other things), and the attention system focuses your brain.[8] This response kicks in whenever the situation we're in starts to feel unpredictable. We have an in-built capacity to judge whether or not the immediate future is likely to be dangerous, and, if everything looks predictable and calm, we switch off our energy-consuming attention and arousal systems. Whenever something surprising happens, however, our attention and arousal states rise very quickly—and initially, we always expect the worst. As David Huron puts it in his excellent book *Sweet Anticipation:*

> Since surprise represents a biological failure to anticipate the future, all surprises are initially assessed as threatening or dangerous.[9]

But surely some surprises are pleasant. An unexpected gift? A phone call from a good friend?

Yes, thankfully there are pleasant as well as unpleasant surprises—but the nice ones are judged to be pleasant only after an initial "threat" response, which is usually so short-lived that it doesn't register with your consciousness.

Imagine you're a child reading in bed. The bedroom door clicks open. It's your mother coming to kiss you good night. The click comes as a surprise, even though your mother comes in every evening at about this time. For the first one sixth of a second or so your mind is running around ringing alarm bells and insisting that the wolves are just around the corner. And then, before you've even noticed that you are thinking negative thoughts, you realize that the whole click/kiss thing is a positive event. And this is not just a childish response. Twenty-five years later you're sitting in bed reading the Sunday papers and the bedroom door clicks—and it's your partner with a tray of tea and toast. Once again, for the first sixth of a second it's all wolves, earthquakes, and Viking raiders, until these negative unconscious responses are washed away by a proper realization of what's going on.

At the moment the click happens, an immediate "action stations" signal is sent off to the fear center of the brain (the amygdala). At the same time, a "what's going on?" message is sent off that reaches the amygdala about one sixth of a second later because, on the way, it passes through the main section of the brain, where decisions are made about what's happening: Does the fear signal need to be escalated or should it be turned off?

No matter how often the door clicks, we *always* have an initial negative response that is then either backed up by a conscious negative response (This really is an earthquake!) or swept away by a positive one (Ah! Tea and toast!). The reason for our inability to learn that a bedroom door clicking in Manchester is unlikely to be a danger signal is that the human race would never have survived if it could un-learn the fight-or-flight response to unexpected sounds (or unexpected touches or tastes or visual events). We have to put up with having brains that always overreact initially because the alternative would mean underreacting on the rare occasions when something dangerous really is going on. Our over-responsive brain might make you a little too jumpy, but an under-responsive brain would eventually kill you.

The fact that our brain responds subconsciously to any unexpected change in what we are listening to has been confirmed by (EEG) brain sensors. For example, even the brains of four-month-old infants give off a surprise signal if they are listening to two alternating notes and the pattern is changed. (If the notes are going ping pong ping pong ping pong, the babies register surprise if they hear two pings next to each other.)[10] Tests like this on adults and infants have shown that we are alerted by any oddity in a pattern of sounds. Unexpected changes in pitch, timbre, loudness, and even the direction the sound is coming from surprise us to some extent. And the fact that we are equally surprised by increases *or* reductions in loudness (or even sudden silences) shows that it's the change in pattern which is important. A pattern of sounds makes you expect a continuation of that pattern, and if your expectations are violated, your brain starts paying attention. Obviously, small deviations generate minor responses and big pattern changes are more startling.

This surprise response is central to our enjoyment of music. When we listen to a familiar type of music, we are presented with a stream of sounds that are generally fairly predictable and only occasionally surprising. The surprises can be based on note choice or note timing (among other things), in either the melody or the accompaniment. When the music is predictable, we get a warm sense of satisfaction from understanding what's going on—which is one source of musical pleasure. If something surprising happens, we always experience an initial negative response that will almost always be followed by a wash of relief that the surprise was merely musical. This negative-positive change magnifies the positive effect you feel. The magnification effect can be illustrated by the following analogy.

Every month you and your family visit your grandmother. Understanding what it's like to be a perpetually broke teenager, your grandma has started to secretly slip a $20 bill into your inside jacket pocket before she hands it back to you as you leave. After one visit you feel for the $20, and it's not there. Damn! Grandma must have forgotten.

But five minutes later you find that she's put the money into a different pocket—and it's not $20, it's $40. Excellent! In this case your expectation was $20, and simply finding $40 in the usual pocket would have been great. But finding $40 after thinking you'd "lost" $20 feels even better. The initial negative has amplified the eventual positive.

The subconscious relief you feel as your brain kicks in to calm down its fear center is one reason why we feel pleasure—sometimes extreme pleasure—when something unexpected happens in a piece of music. If you are listening to a discordant, jarring section, you will be predicting more of the same in the near future. But if the music surprises you by turning all tuneful and agreeable, the pleasant effect will be amplified by the surprise involved.[11] The fact that the initial negative response is biological and can't be turned off is the reason why a certain passage of music can have the same "goose-bumpy" effect even when we know it well. Another contributory factor to our pleasure is the expectation that the marvelous goose-bumpy bit is just about to happen.*

The tingle factor

Sometimes your emotional reaction to music can produce a strong physical reaction. Professor John Sloboda has identified three main physical reactions to music—which can be experienced singly or in combination.[12] They are:

1. A "lump in the throat" feeling, sometimes accompanied by tears (often these are tears of happiness or relief).
2. A tingling skin sensation—goose bumps as the hair on your skin stands on end—sometimes accompanied by shivers down the spine.
3. An increased heart rate, sometimes combined with a sinking feeling in your abdomen.

This level of emotional engagement happens only occasionally, and it's more likely to happen with music you are familiar with

* One of my more recently discovered goose bump moments is in a track, "The Birds," by the band Elbow (on the album *Build a Rocket Boys!*)—the synthesizer entry three minutes, twenty-five seconds into the track.

and when you are listening carefully—not just listening as you fill the dishwasher. Although it's difficult to pin down what causes this type of thrill reaction, there is evidence that it's often linked to an unusual change in the music—a violation of your expectations of what this sort of music would normally do at that point. Obviously if you are familiar with the piece of music involved, you know that the unusual change is coming, but it still has the capacity to thrill you as long as you're in the right mood. Of course, some of us are more prone to emotional musical thrills than others. Looking into this, several psychological studies have suggested that people with the personality trait "openness to experience"* are more likely than others to feel intense emotions as a result of listening to music.[13]

Although there are no firm and fast rules (as usual with music), the three types of physical/emotional reaction have been linked to the following musical techniques.[14]

The "lump in the throat" feeling with occasional tearfulness can be caused by various musical effects including something musicians call a "melodic appoggiatura." In plain language, this involves singing or playing a melody note that doesn't fit with the accompanying harmony, and then shifting up or down a step in pitch to a much more suitable note. This release of tension is the musical equivalent of going "phew" after almost missing your train.

Shivers and goose pimples are usually caused by changes in the harmony rather than the melody. The music may be tootling along normally and then, without warning, the composer starts using different notes for the chords that provide the harmony. It's a bit like watching an outdoor sporting event through orange-tinted

---

* People who have a high "openness to experience" rating tend to have above-average levels of active imagination, aesthetic sensitivity, attentiveness to inner feelings, preference for variety, and intellectual curiosity.

sunglasses and suddenly changing to green-tinted ones. A lot of what's going on hasn't changed—but something fundamental is different.

Increases in heart rate are, as you might expect, linked to the timing of the music. For example, your heart rate can increase if the music you're listening to goes from a steady, straightforward rhythm to a syncopated one. "Syncopated" just means that the rhythm puts emphasis in unusual places. Instead of the usual:

One and Two and Three and Four and...

we might get, for example:

One and Two and Three and Four and...

Another way for the composer to accelerate your heart rate is to build up your expectations for a climax and then provide it slightly early.

You might have noticed that all of these big emotional effects are based on building up expectations and then violating them (but not too much), a principle that also works for storytelling and jokes. We like things to take a path that makes sense, but we don't want it to be too predictable. One big difference between music and storytelling or jokes (as far as adults are concerned) is that repetition improves our attachment to a piece of music. We generally experience the strongest emotions from music we are familiar with.[15] In the case of jokes and stories, although we all have our favorite film quotes and snippets from TV sitcoms, we usually get the maximum impact the first time we hear them.

Not very James Bond, are we? We like musical surprises—but only if they're not *too* surprising—and we prefer surprises we've heard before.

CHAPTER 5
Music as Medicine

Modern music therapy began as a simple attempt to cheer up American soldiers who had been wounded or traumatized in World War II by holding concerts in veterans' hospitals. The medical staff soon noticed that the music seemed to be having a surprisingly positive effect on the physical and mental condition of the patients, so hospitals began hiring musicians to give their patients more regular exposure to music. Eventually everyone realized that the positive effects would be further enhanced if the musicians were trained as therapists, and so the first music therapy degree course was started in 1944.

The aim of music therapy is to use our subconscious responses to music to assist our recovery from a medical or psychological condition. Over the past few decades music therapy has been found to be effective in a wide range of applications. Here are a few examples of how it works.

Music therapy for depression and stress

One of the big problems with being depressed is that you're too depressed to see any way out of your depression. Depression creates a vicious cycle: negative thoughts generate negative moods, which encourage negative thoughts, which generate negative moods, which generate...

Music has the power to break this cycle. This is not just happy-clappy mumbo-jumbo. Listening to pleasant music causes your serotonin and dopamine levels to rise, resulting in a genuine, positive mood change.[1] It's been found that music even generates increased levels of dopamine in stressed rats.

Once a mood change has been achieved, depressed patients are more receptive to positive thoughts and guidance to help them maintain a more positive view of themselves and the world around them.[2] I suspect this is even true of the rats—but I'm not sure we want rats with a well-balanced worldview.

Depression is often linked to stress, so psychologist Suzanne Hanser decided to see if music-related activities could reduce stress and thereby reduce depression.[3] She contacted thirty elderly people who were suffering from depression and, after getting them to agree not to take any antidepressant drugs for the duration of the eight-week test, divided them into three groups of ten.

The first group underwent personal training (at home, one hour a week for eight weeks) in music-related stress-reduction techniques, including movement to music, relaxation to music, using music to become energized, and using music to induce visual imagery. This group chose their own music for these exercises.

The second group received a diluted version of the first group's treatment. They were given written instructions on what to do and suggestions of what music to use. They also got a phone call once a week from Suzanne in order to deal with any questions and to give them advice.

The third group got nothing—just the promise that they would be able to begin the therapy at the end of the eight weeks. "Pah!" they must have thought. "... Typical!"

All three groups were tested for depression and stress levels at the beginning and end of the eight-week period, and the results were impressive. All three groups had been (mildly) clinically depressed and stressed at the beginning of the period, and the group who had no therapy were in much the same state at the end. The two music therapy groups, by contrast, had improved so much that their depression and anxiety scores were closer to those of non-depressed people than to their own previous scores. What's more, the improvement was long-lasting. Nine months later the depression/anxiety states of the music therapy groups hadn't changed much, and some people were still improving.

For one of the subjects the music therapy was particularly successful because the music itself was important. Judy was sixty-nine years old when the study began, and having a dreadful time dealing with the recent death of her husband, Bernard. Bernard had been a clarinet player who loved big band music. Suzanne Hanser suggested that Judy use Bernard's old big band records as her therapeutic music, and within a week there was a marked change in Judy's attitude toward life. Nine months after the experiment, Judy wrote to Suzanne:

> Before I took out all those records, I spent every day thinking that I would never see my husband again. I loved him so much. I couldn't bear to be without him. But now, when I put on his music, it feels like he is with me. I think about all the great times we had and how much we shared. I love the records and I love listening to his records. Thank you for giving me back my husband.[4]

I think you'll agree that this is pretty much an unbeatable result.

Music therapy and pain

Considering how widespread it is and how unpleasant it can be, we know surprisingly little about pain. Pain is not a thing—it's a perception. The same injury can cause vastly different levels of pain in different people depending on a number of factors, including how busy they are and what mood they are in. If you stub your toe during a relaxed Sunday afternoon, you might feel a lot of pain. If you do the same thing as you are running to stop a small child from walking in front of a bus, you might not notice the pain at all.

Fortunately, because pain is a perception—a feeling—rather than something more concrete, music can help to minimize it. Music, as we've seen, can reduce stress, relax you, improve your mood, and focus your attention—all factors that can help reduce the pain you feel. In addition, the fact that your brain is having to process the music is a distractor—like seeing a child in danger—which interferes with the "Bloody hell! That hurts!" signal.[5] Music has been found to be very useful in dealing with temporary pain such as dental treatment and headaches, particularly if the patient chooses the music and controls how loud it is. Interestingly, the music works best if the patient has been *told* that it will reduce the pain. If patients believe that they have some control over a method of pain reduction, the belief itself helps to reduce the pain.[6]

One study of the effects of music on pain involved asking volunteers to keep their hands in very cold water for as long as they could stand. Participants who chose their own music could keep their hands in the cold water for longer than people who listened to white noise or random relaxation music. Once more it seems that choosing the music made the volunteers feel more in control of the situation and this helped them cope with the pain for a longer time. Women who performed this test after choosing their own music not only coped longer with the pain but also felt that the pain was less intense; men coped longer but felt the same intensity.[7]

Music and speech therapy

When I was a child I had a guilty secret. (Now that I'm in my fifties I probably have several guilty secrets, but my memory is so bad

I can't remember them. Hurrah!) When I was nine years old, my secret was this: although I was an avid reader and therefore knew all the letters in the alphabet, I didn't know what order the letters were in. I knew as far as A, B, C, D, E. But after that it was all a bit vague. Did M come before Q? Where did T fit in? I can't say I lost much sleep over it, but it did trouble me that everybody else seemed to know. Then, when I was ten, someone sang an alphabet song to me, and I could remember the whole thing after about a dozen repetitions. The world took on a new luster; doors of opportunity opened up before me: I could even become a librarian when I grew up if I wanted to. I still rely on that alphabet song to this day. (It's not the one most people know. It must be some sort of Manchester 1960s hybrid.) I don't know if you have to do this, but if I'm looking through an index of any sort, I have to sing a bit of the song under my breath. If I want to know exactly what letter comes after L, I have to sing a full musical phrase from the song—"IJKLM." So I don't have a "proper" memory of the letters of the alphabet; I just know a song that can deliver the information on request.

Tunes are very handy for this kind of rote learning. It's quite easy, for example, to teach kids to sing a song in a foreign language. As long as there is a melody involved, they can be trained to pronounce the words accurately even though they haven't a clue what they are singing about. I'm sure many of you will remember singing "Frère Jacques" at school. You can probably still sing it, but I wonder how many of us kids knew what "Sonnez les matines" meant?* I think I was in my thirties before I found out. But it didn't matter. At the age of six I could still pronounce fourteen seconds' worth of understandable French words—which

* For those of you who sang this song at school without ever being told what it's about: Frère Jacques is a monk and we are singing to him to wake him up so he can ring (*sonnez*) the bell for morning prayers (*les matines*).

were just random sounds as far as I was concerned. This sort of memory feat would be incredibly difficult without the assistance of a tune.

This musical memory trick can be used to help some patients who lose the ability to speak as a result of a stroke or brain damage. In his excellent book *Musicophilia,* Oliver Sacks tells the story of a patient, Samuel S., who had been left totally without speech as the result of a stroke.[8] Although he understood what other people were saying, Samuel could not pronounce a single word, even after two years of speech therapy. His brain simply didn't know how to make his vocal equipment produce words anymore.

The situation was considered hopeless, but one day a music therapist overheard him singing a fragment of the song "Ol' Man River." Samuel still couldn't speak, and he could sing only a couple of words, but it was a start. The music therapist began regular singing lessons with him, and soon he could sing all the words of "Ol' Man River" and lots of other songs he had learned when he was younger. The singing kick-started his speaking abilities, and within a couple of months he could produce appropriate, though short, responses to questions. For example, if someone asked him how his weekend had gone, he could reply, "Had a great time" or "Saw the kids." Music has been helpful in this way for a lot of patients for whom standard speech therapy didn't work.

Music therapy, blood pressure, and heart attacks

Coronary heart disease is the most common cause of death in a lot of countries, including the United States and the UK. What generally happens is that prolonged stress causes a long-term increase in blood pressure, called hypertension, which often results in heart disease, heart attacks, and strokes. Experiments have shown that music therapy (music with relaxation/imagery exercises), combined with standard patient support, produces noticeable improvements in blood pressure, anxiety levels, and general well-being. As Suzanne Hanser puts it:

> Individuals with cardiac conditions were able to take charge over the stress in their lives when they learned to recognize how music changed their heart rate and blood pressure.[9]

Biofeedback devices, which monitor your blood pressure and pulse rate, can be very useful in this context. You can put some of your favorite relaxing music on the stereo and watch your pulse rate and blood pressure fall, and train yourself to relax.

Music therapy and Parkinson's disease

One of the main symptoms of Parkinson's disease is that your physical movement becomes jerky, rather like a verbal stutter. It's well known that people who have a verbal stutter generally lose it if they sing their words, but music can also eliminate or reduce Parkinsonian movement stutter. In the words of Oliver Sacks:

> Parkinsonian stutter can respond beautifully to the rhythm and flow of music, as long as the music is of the "right" kind—and the right kind is unique for every patient.[10]

The "right" kind of music often needs to have a well-defined beat, but if the beat is too dominant, the patients can become enslaved and locked into it. The helpful effect of the music usually ends as soon as the music stops, so the modern proliferation of portable music players has been a great help to a lot of Parkinson's sufferers.

Music, general well-being, and sleep

Music lowers stress levels, and this can have a positive effect on your entire immune system. For example, several studies have shown that music improves the level of the antibody immunoglobulin A in your saliva, which is a direct measure of the ability of your respiratory system to fight off infection.[11] Psychologist Shabbir Rana has also found a direct link between the number of hours people spend listening to music and their scores on the General Health Questionnaire,* which is a test of your psychological well-being.[12]

And there is good news for a lot of those people who have trouble sleeping. As most of us know, getting a decent amount of high-quality sleep is extremely important to your quality of life. Sleep disorders can lead to fatigue, anxiety, depression, and poor daytime performance in both physical and mental tasks. Drugs can help, but they can also have negative effects on your daily life. Fortunately researchers have found that simply playing relaxing music at bedtime can alleviate sleep disorders for many people. Relaxing music reduces the amount of the stress hormone noradrenaline in your system, thereby reducing your level of vigilance and arousal and allowing you to sleep better.

If you're a scientist working in the area of sleep problems and you need a reliable measure of how badly people sleep, you need look no further than . . . Pittsburgh. The Pittsburgh Sleep Quality Index (PSQI) is a four-page questionnaire about your sleeping patterns that gives you a score telling you how good a sleeper you are. If your score is lower than five, then you are a normal sleeper. (If you score one or less, you might actually be dead.)

---

* The questionnaire is available in various forms and asks from twelve to sixty questions about how you've been feeling over the past few weeks.

Armed with the PSQI, psychologist Laszlo Harmat and his colleagues gathered together ninety-four students with sleeping problems and divided them up into three groups. The students were all between nineteen and twenty-eight years old. Before the tests began, all the subjects completed the PSQI questionnaire, and on average they scored six and a half: they were all poor sleepers. One group was given relaxing classical music to listen to at bedtime, a second group was supplied with an audio book, and the third group received nothing. Those with the music or audiobooks were asked to play them every night for forty-five minutes just before they went to bed.

After three weeks of bedtime listening, the average score in the music listening group dropped to just over three: they were now good sleepers—or at least most of them were. Of the thirty-five people in this group, thirty of them became good sleepers and the other five remained poor sleepers. Listening to audiobooks helped far fewer people: only nine out of thirty became good sleepers. The same students were rated as to how depressed they were before and after the three-week test. The depressive symptoms of the music-listening group decreased substantially during the test, but the audiobooks didn't have the same effect.[13]

Music doesn't just help young people, though; it also helps older people sleep better. In 2003 researchers Hui-Ling Lai and Marion Good carried out a similar study on people aged between sixty and eighty-three years old.[14] For this group the PSQI questionnaire revealed an average score of over ten: they were very bad sleepers.* The researchers handed out music tapes that were forty-five minutes long and asked the participants to listen to them after they had gotten into bed. (It normally takes an adult between thir-

* About half of the people in this age group have occasional sleeping problems, and a substantial proportion of them take hypnotic drugs to help them sleep (although none of the participants took these drugs during the test).

teen and thirty-five minutes to drop off to sleep.) Once again, the bedtime music worked its magic, although this time on a smaller proportion of the people involved (possibly because the sleeping problems of this older group were more deeply embedded). Half of the music-listening group dropped below a PSQI rating of five and became good sleepers.

If you want to try relaxing bedtime music for yourself, you'll find that there are plenty of "the most relaxing classical/jazz/blues music in the world/galaxy/universe" albums available, and of course you can also make up your own playlists. When you play them, it's important to get the volume just right: too low and it's irritating, too loud and you can't sleep. And make sure that the final piece is one that fades out; otherwise you'll be woken by the sudden silence. (One of our natural reflexes is to go on guard if it suddenly gets quiet.) Also, if, after a couple of months, you get bored to tears with "Air on the G String" and "Für Elise," you could always do what I do and play albums of lute music.*

Music therapy: the future

The evidence is pretty overwhelming that listening to music you enjoy is good for you if you have either a physical or a mental disorder of any sort. The choice of music is important and, as we've seen, the patient must do the choosing. But what if you are suddenly struck down by a medical condition that prevents you from making choices?

This worry has led to the invention of the Advance Music Directive,[15] also known as the "music living will." With the assistance of a music therapist, you create playlists to be stored by someone you trust. Then, if anything dreadful happens to you, the

---

* My present favorite is an album called *Cantabile* by Nigel North.

appropriate playlist (calming, energizing, etc.) will be played to assist your recovery.

All this reminds me of a conversation I sometimes have with my friends in which we discuss the music we want to be played at our funerals.* The only difference is that the sickness playlists are much longer, and you have to be *much* more careful about your choices. You don't want to be lying in a coma thinking "Oh no!—not 'Seasons in the Sun' again!"

So if you have a beloved friend or relative in some horrible bed-ridden state, make a playlist or two—it's bound to do some good. If he is conscious and able to communicate, you could get him to help you compile the playlists (one for relaxing/sleeping, one for positivity). If he is unconscious, try putting the playlists together yourself—but remember two things: (1) The playlist should consist of music you imagine *he* would choose—not just music you would choose for yourself; (2) Never, under any circumstances, include "Seasons in the Sun" or "Chirpy Chirpy Cheep Cheep."

* I'm going for "Blue, Red and Grey" by the Who.

CHAPTER 6

Does Music Make You More Intelligent?

The Mozart effect

In 1993 a team of psychologists led by Frances Rauscher published a paper called "Music and Spatial Task Performance" in the highly prestigious scientific journal *Nature*.[1] The research they carried out was designed to find out how well students performed in a particular intelligence test after spending ten minutes doing one of three things: some of the students spent the time listening to relaxation instructions, some just sat around in silence, and the final group listened to ten minutes of Mozart's piano music. Then they all took the same test.

The results showed that the group who listened to Mozart performed better than the ones who listened to relaxation instructions or did nothing. The enhancement in performance was the equivalent of adding eight or nine points to their IQ—a useful improve ment.

The test involved looking at some drawings of a piece of paper being folded and cut and then predicting what the paper would look like if it was unfolded. Here's the sort of thing they were tested on: Can you see which of shapes A–E you'd get if you folded and cut the paper as shown on the next page? (You'll find the answer at the end of the chapter.)

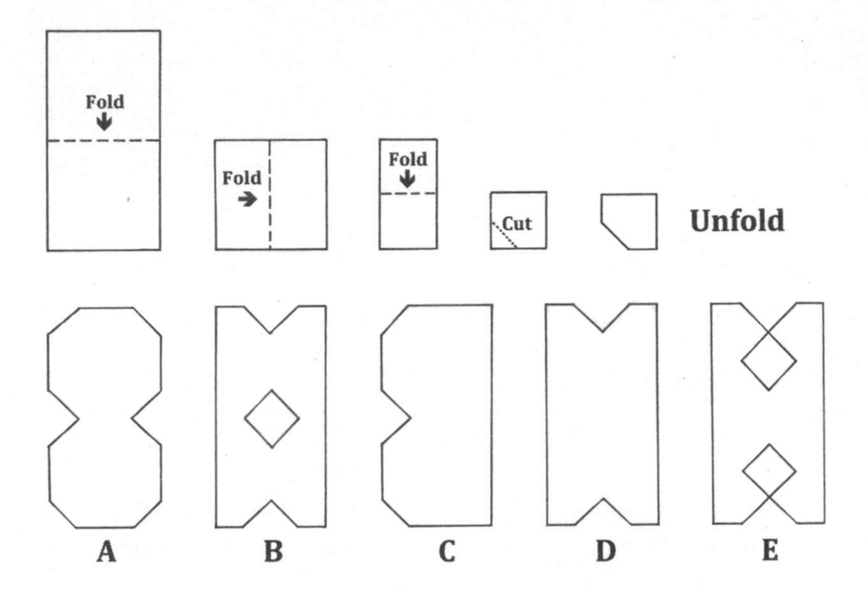

The problem is what happened next.

If Dr. Rauscher and her team had published their results in an obscure psychology journal, her colleagues in the field would have read it, added it to the pile of stuff we know about how brains work, and carried on diligently adding to the pile. But *Nature* is big league. It's where a lot of the really interesting, world-changing science is published, which means that a lot of newspapers employ an intellectual (or at least someone who wears glasses) to read it from cover to cover every month, looking for stories about cancer cures and trousers that never need ironing and so on.

After the Rauscher results were recycled by these bespectacled newshounds, the media were full of stories about classical music increasing your IQ. The Mozart effect was born. Rauscher never said that listening to Mozart increased general IQ and has explained this to the short-sighted members of the press several times, to no avail. She was only talking about specific skills, having to do with how we visualize the outcomes of certain actions.

Of course, no one listened. The idea that classical music made you brainier was far too sexy to be quashed by facts.

Before long all sorts of people were jumping on the bandwagon. The UK radio station Classic FM brought out a best-selling CD called *Music for Babies*. Back in the States, New Hampshire was handing out classical music CDs to new mothers, Florida passed a bill insisting that state-funded day care centers should play classical music, and Texas prisons played symphonies to inmates,[2] which undoubtedly led to a lot of stabbings in the showers as criminal overlords fell out about the respective merits of Rachmaninoff and Sibelius. By the end of the 1990s, surveys carried out by psychologists Adrian North and David Hargreaves in California and Arizona found that four out of five people were familiar with the Mozart effect. Amadeus would have been delighted.

Since then a lot of psychologists have looked into the possibility of there being a relationship between IQ and listening to Mozart. In 2010 a review of thirty-nine investigations into the subject (involving more than three thousand participants) finally confirmed that there actually is a Mozart effect—but it has nothing to do with Mozart.[3]

Psychologist E. Glenn Schellenberg and his colleagues from the University of Toronto have done a lot of work in this area and, in one of their experiments, decided to swap Mozart's music for an audio recording of a Stephen King novel. This might seem like a strange exchange, but Professor Schellenberg and his team were working on the idea that maybe the music itself wasn't important. Maybe just listening to anything enjoyable before doing an intellectual test put you in a good mood and made you perform better. The people they were testing were university students who, I think we can all agree, are just as likely to enjoy a Stephen King story as a two-hundred-year-old piece of piano music. Lo and behold, the students reliably did better on the same "what will this cut paper look like when I unfold it?" test if they listened to either the music or the story beforehand rather than sitting in silence. After the test the students were asked if they preferred the story or the music. Those who preferred the story got their best score after the story and, similarly, the music brought out the best scores in the music lovers.[4]

So now we don't have to try and work out how Mozart's music manages to magically reweave your brain patterns to make them more effective. The "Mozart effect" is just an example of something that is quite well known—that being in a positive frame of mind improves your intellectual performance.[5]

This "positive frame of mind" is a combination of being in a good mood and being moderately aroused. "Aroused" in this context means the opposite of bored. If you are under-aroused (bored or sleepy), your brain isn't ready to do much work, and you will score badly if someone suddenly hands you an IQ test. On the other hand, you can be over-aroused—in an excitable or panicky state—which will also make you perform badly. You will achieve your highest score if you start the test in a moderately aroused state, which is how you'll find yourself after listening to a story or a piece of music. On top of this, the more you enjoyed the story or music, the better your mood will be—and a good mood can also improve your score. When you are in a good mood, your dopamine levels rise, and this is thought to improve the flexibility of your thinking processes, enhancing your ability to solve problems and make decisions.[6]

Pursuing this idea, the Toronto team started trying all sorts of music before the test to see what would happen, and pretty soon they had discovered the Schubert effect, which was, unsurprisingly, exactly the same as the Mozart effect. Other composers also worked well, but the team forecast that there *wouldn't* be an Albinoni effect. Tomaso Albinoni is one of music's one-hit wonders.

You've probably heard of his Adagio,* but this piece is so miserable and slow that it's unlikely to raise your arousal level, or put you in a good mood. Sure enough, when this music was used before the fold/cut test, no Albinoni effect emerged.[7]

The Mozart piece used in the original test was in a major key and was moderately fast in tempo, giving it a chirpy, happy feel. The Toronto team tried playing this piece to their participants at a slower speed and also had an altered version of it made in which it was transposed into a minor key (since, as we've learned, minor keys tend to evoke sadder emotions). Using these different versions, they confirmed that the faster music was more effective at arousing people and major key music produced better moods—so faster, major key music led to optimum performance on the test.

By this time the team had run out of test victims in Toronto and decided to invade the UK, where they roped in the BBC to help them test nearly eight thousand schoolchildren at the same time. In each of about two hundred schools, the ten- and eleven-year-olds were divided into three groups and herded off into three rooms, each of which contained a radio. One group listened to the pop band Blur on BBC Radio 1, a second group listened to Mozart on BBC Radio 3, and the third group listened to psychologist Susan Hallam discussing the experiment they were taking part in. After this listening exercise the kids performed a couple of tests on their spatial abilities. As we might expect from the good mood/arousal theory, the children performed best after the listening event that had stimulated them most and that they enjoyed most—so now we have the Blur effect.[8]

The upshot of all this is that it doesn't matter what you listen

* Actually, Albinoni might be music's only *no*-hit wonder. There's a good chance that "Albinoni's" Adagio was actually written by Italian musicologist-composer Remo Giazotto in the 1950s.

to—Mozart, Blur, or Stephen King stories—before you head into your next mentally challenging activity. Listening to anything mildly stimulating that you enjoy will temporarily improve your brain's performance.

These tests were all looking at what happens when both your arousal level and your mood are improved by listening to something. But although they often rise or fall together, your arousal level and your mood are not always linked. You can be happy and sleepy or happy and excited. Your mood is linked to how much dopamine you have sloshing around in your brain at any given moment, and your arousal is linked to an entirely different chemical called norepinephrine.

In order to separate out the effects of music, mood, and arousal level on mental performance, psychologists Gabriela Ilie and William Forde Thompson carried out an experiment on several groups of people in 2011. They asked subjects to complete mental tasks and creativity tests after listening to seven minutes of recorded classical piano music. Some of the listeners heard the music played loud, fast, and high on the piano keyboard; some heard it quiet, slow, and low-pitched; and other groups heard different combinations of the loudness, speed, and pitch height. The tempo of the music had the biggest effect on arousal (the faster music was the most arousing), and the pitch was more important as far as mood was concerned (the higher-pitched recording made people happiest). So the music produced groups with differing levels of mood enhancement and arousal.

The creativity tests the listeners had to do are known as Dunker's candle problem and Maier's two-string problem. For the candle problem, you are given a box of thumbtacks, a book of matches, and, of course, a candle. The task is to use what you've been given to attach the candle to the wall in such a way that it will burn without dripping wax on the floor.

(If you're in the mood you might want to try to work out a solution before you read the next line...)

A typical creative solution to the problem is to throw away most of the tacks, use two of them to pin the box to the wall and then rest the candle on the box.

Maier's two-string problem is a lot more irritating, especially if you're not feeling particularly creative that day. You are presented with two lengths of string hanging from the ceiling and asked to tie the two dangling ends together. The irritating bit is that if you hold onto one dangling end you can't reach the other one. The only equipment you are given to help you complete the task is a pair of scissors and a chair.

(Once again you might want to think about how you'd solve this before you read on...)

I'm sure there are some people out there who've thought of some way of using the chair, but you don't need it. Just tie the scissors to the end of one of the strings and start them swinging. Then, holding onto the end of the other string, you grab the scissors as they swing toward you. Once you have hold of both strings it's easy to tie them together.

The final test the listeners were asked to complete was designed to require very little creativity. Instead it demanded high speed in doing a simple mental task. On a computer screen filled with 408 geometric characters, the listeners were asked to find and click on a particular type of character that appeared on the page fifty-eight times.

And the results? Well, if the music had resulted in high arousal without much mood enhancement, then the people involved tended to do very well at the high-speed simple task and not so well on the candle and string problems. By contrast, those who had experienced high levels of mood enhancement without much stimulation tended to do better at the creativity tasks but were slower at the simple character identification test. So the two conclusions of this experiment are: (1) improvements in mood help you think more creatively, and (2) increased stimulation makes you faster at easy thinking tasks.[9] As I said earlier, your mood and stimulation level often rise together when you're listening to music, giving you both benefits at once.

So far we've looked at the results of listening to music *before* you carry out a mental task. But what about background music playing *while* you are doing something that requires thought and focus—like studying, or filling in a tax return?

Does background music help you think?

Because it's of interest to nearly everyone, from students to call center managers, hordes of psychologists have looked into whether or not background music is a help or a hindrance when you are trying to carry out an intellectual task. They've looked at our performance in reading comprehension, our memory of what we've just read, and our ability to do math. So far the result of all this research is a resounding "It depends."

As we saw in the previous section, music, particularly music you like, will put you in a good mood—and good moods improve your thinking. On the other hand, you have only a limited amount of brain power, and some of it will be busy processing the music, leaving you with less to channel into the work. This processing of the background music is, of course, obligatory, because you can't close your ears.

Whether background music is useful or intrusive depends, among other things, on what you are trying to do and where you are trying to do it. For example, it has been shown that surgeons learning how to carry out a new operation find the task more difficult if background music is playing.[10] By contrast, students trying to read news items off a pocket computer in a noisy university cafeteria found that music helped them read faster and remember more of what they read.[11] In the case of the surgeons, the alternative to the music was a generally calm, quiet environment—so the music was a distraction that used up some of their thinking capacity. For the students, the music (fast-tempo classical) was a welcome relief from the noise of clattering dishes and the voices of friends moaning about exam results and/or passing judgment on so-and-so's new haircut. The music masked these noises and so *reduced* distraction.

If you are in an otherwise quiet environment, the distraction level of the background music will be highest for loud, busy (fast, jumpy) music. Loud music is more insistent that your brain deal with it, and the busier the music is, the more brain power is required to process it. There is also experimental evidence that music with vocals is more distracting than instrumental music—a phenomenon I have often noticed when trying to get some work done.

So if you are going to have music playing while you try to untangle your tax situation or help your daughter with her science homework, make it calm, quiet instrumental stuff. This should bring you the twin benefits of improving your mood and masking background noise (the neighbor's vacuum cleaner, the sobbing of your daughter) without overly distracting you.

Does learning a musical instrument make you more intelligent?

There is a lot of evidence that musically trained children perform better intellectually and academically than their friends who haven't been musically trained. But it has also been shown that kids with higher intelligence levels are more likely to take music lessons.

So is musical proficiency a cause or merely a symptom of higher intelligence?

When it comes to human behavior, cause-and-effect questions

are tricky. You'd think, for example, that happiness caused you to smile rather than the other way around. But it's not that simple. Smiling and happiness are intertwined; they feed off each other. Just the mechanical act of smiling actually makes you more cheerful. This sounds like mumbo jumbo, but all you need to prove this point is a bunch of people and some pencils. You ask half of the people to put the pencil between their lips sideways and bite gently down on it—the way a horse holds a bit in its mouth. The other group put the end of the pencil in their mouth and purse their lips around it so the sharp end is pointing forward. Your subjects won't realize it, but you have forced one group to smile and the other group to frown. Tests have shown that if you then give these groups something amusing to look at, like Gary Larson cartoons, the "forced smilers" will find the jokes funnier than the "forced frowners."[12] So your mum's advice to "put a smile on your face—it'll make you feel happier" is actually true.

As you can imagine, if the causes and effects of something as apparently simple as smiling are this entangled, finding out if musical proficiency is a symptom or a cause of higher intelligence in kids is an incredibly difficult task.

Let's look at the facts that various teams of psychologists have unearthed in the past couple of decades.

Musically trained people are better listeners. They are generally better at detecting, for example, subtle changes in pitch in the final word of a sentence. This ability makes some, but not all, musically trained adults and children slightly better at identifying nuances of emotion being expressed by other people.[13]

Musically trained people have a better memory for things they have heard—whether the subject matter is music or words.

Musically trained young children perform better on tests of language ability; for example, they add words to their vocabulary more rapidly.[14]

- Musically trained people have better visuospatial skills—the skills that allow you to recognize shapes and distances and make sense of what you see around you. Musicians, for example, have an easier time than non-musicians seeing shapes hidden in complicated line drawings and perform better on "what's the difference between these two pictures?" tests.[15] One odd outcome of the improvement in visuospatial skills is that musicians divide a horizontal line in two differently from non-musicians. When asked to identify the midpoint of a horizontal line, non-musicians tend to put their mark to the left of center. Musicians tend to put their mark nearer the center point—but slightly off to the right.

Many people believe that there is a link between music skills and mathematical ability. Psychologists have looked into this, but the results suggest that the connection between the two is either tiny or nonexistent. Some researchers have shown a small positive link between music training and mathematical skill, but a study of more than seven thousand fifteen- and sixteen-year-olds carried out in 2009 showed no relation between the two.[16]

Another study looked at a possible music-math relationship from the other direction, by trying to find out if having a high level of mathematical skill had any correlation with musical ability. Adult members of the American Mathematical Association were compared with members of the Modern Languages Association, and both groups were found to be equally musical.[17] So I'm afraid that the widespread idea that musical ability is often accompanied by math skills is just a myth.

The fact that musical training is linked to better listening skills, language skills, and visuospatial skills is an indication of improved brain function, but does this mean it can improve your general intelligence, as measured by your IQ?

Testing in this area shows that the IQs of children undergoing musical training are generally higher than those of other children by between ten and fifteen points. It would be tempting to attribute this to the fact that the brighter kids are the ones who tend to have musical training. Professor Glenn Schellenberg, however, has carried out an experiment which suggests that music training really does *cause* a slight increase in IQ.[18] He divided 144 six-year-olds into three groups. One group were given a year of music lessons (on top of their normal school lessons), the second group were given drama lessons, and the third group were given no extra lessons. The children completed IQ tests at the beginning and end of the year—and music lessons clearly raised the IQ of the music group by about three points compared to the others. So the effect is small, but there's evidence that it does exist. (One interesting side issue is that although the drama students didn't experience an improvement in IQ compared to the kids who did no extra lessons, their social skills were the best of all the groups by the end of the year.)

So, like the connection between cheerfulness and smiling, it seems that the link between IQ and music lessons goes both ways. Brighter kids are more likely to have music lessons, and having music lessons tends to make you brighter.

The answer to the fold/cut/unfold problem at the begining of this chapter is B.

CHAPTER 7

From *Psycho* to *Star Wars:* The Power of Movie Music

Jealousy is not a pleasant trait—but I have to admit to being jealous of my namesake, John Powell. For the half dozen or so of you who haven't heard of him, he's the guy who wrote the music for the *Bourne* trilogy, *How to Train Your Dragon,* and lots of other excellent stuff.* I've always thought that being a film composer must be a great life—hanging out with the rich and famous, *being* one of the rich and famous. There is one downside to the job, however, and that's the amount of work involved. Movie music is usually written to tight deadlines and needs to be carefully coordinated to the action, as well as being generally unobtrusive. It also needs to be beautiful in its own right if you want to start collecting Oscars. This is a pretty tough confluence of requirements—so, John, if you're reading this and need a couple of days off, no problem, just give me a call and I'll come over and fill in for you. I wouldn't be able to do any of the difficult, talented stuff, but I

* Whenever I give a talk about music, I have to begin by saying, "Before we start, I'd better point out that I'm not *the* John Powell—I'm just *a* John Powell." This usually gets a laugh from most of the audience—but there are always a couple of people who look a bit disappointed. The disappointment doesn't last long, however, because within seconds my bodyguards have them forcibly ejected from the auditorium and thrown out onto the street.

could have a go at a "man gets into taxi at airport" scene or "woman looks in handbag for door keys" or something.

Back to reality . . .

In the days of silent films a pianist would accompany the movie, and some of the fancier cinemas even employed a small orchestra. Initially the music was there only to mask the clatter of the projector, but it was soon realized that music could help the story along as well.[1] This led to the development of anthologies of sheet music designed to accompany different sorts of action: desperate running, villainous creeping about, passionate kissing, and so on. This sort of "mood music" became so important to silent films that it was even played during filming, to get the actors into the right frame of mind.[2] When the first talkies were made, the directors assumed that the action and words would be enough, and that music would be irrelevant. Audience responses soon proved them wrong, and since then a musical sound track has been produced for nearly every film.

Music can enhance our enjoyment of a film in many ways. If the action on the screen is ambiguous, the music gives us clues as to what's going on, which is particularly useful right at the beginning of the film, when we are, for example, just watching someone walking down the street. If the music is dramatic and driving, we can guess that we're about to see some action; if the music is happy and daydreamy, then the situation will usually be cheerful—although composers do sometimes give you pleasant music just before a disaster, to increase the shock you experience. Psychological tests in this area have shown that music helps you understand what's going on and makes it easier for you to forecast what sort of thing is going to happen next.[3] Music that evokes associations also has the power to draw your attention to certain visual details. For example, a soldier walking through a crowd of civilians is more easily picked out by the viewer if a solo military trumpet is involved in the music.

Some movie sound tracks become hit albums, which means that

the music was consciously enjoyed by the audience. This is particularly true of the music played during the opening or closing credits of the film (think of all those James Bond songs). Most movie music, however, is designed to be a subconscious influence, and people don't usually notice it. If they are asked to watch a film sequence and discuss it, the audience members almost always think they have drawn their conclusions from purely visual information—even though music may have been doing half the work.[4]

In one experiment, viewers were presented with a five-minute clip from the Hitchcock film *Vertigo*. The clip was shown to different groups of people together with one of three sound tracks: no music, emotional music (Barber's Adagio for Strings), or harsh, unfriendly music (part of Tangerine Dream's *Rubycon*). The clip showed a man (Jimmy Stewart) following a woman (Kim Novak), and the viewers were asked what they thought about the man. When no music was playing, the audience had the opinion that the man was an intelligent, analytical private investigator. With the emotional music playing, they described him as a sensitive, infatuated lover, but the harsh electronic music made him come across as a cold, lonely hit man.[5] These widely different readings of the visual action show just how persuasive music can be.

As we saw earlier, music on its own isn't very good at communicating a story—but when it's used together with film, it can be highly effective. For example:

The film *Witness* is about an Amish boy who witnesses a murder. In one of the early scenes in the film, the boy is taken down to the police station to look through a photo book of possible suspects. When he is left alone for a moment, he spots a framed photo of a police officer and it dawns on him that the officer is the murderer he saw. At this point the background noises of the police station are replaced by the otherworldly music of Maurice Jarre. Thanks to the music and the absence of background noise, we are (subconsciously) alerted to the fact that the photo is significant. The music is doing an important job here because the rest of the plot depends on this identification scene.

Psychologist Annabel Cohen and her colleagues used this clip in a test to find out just how effective the music was.[6] Participants were shown the minute-long clip of the scene at the police station with one of five sound tracks: silence; speech only; sound effects (typing, etc.) only; a combination of speech, sound effects, and music; or the original "music only" version. They were then asked to rate their sense of absorption in the clip—how much they were drawn into it. The results clearly demonstrated the power of the music to engage people with the film. The viewers found that the music alone was more effective even than dialogue in getting the desired message across.

Besides giving us hints as to what's happening, music can provide clues to when and where the action is taking place. For example, a flashback to the 1930s might be accompanied by jazz—played on a suitably dated, crackly radio. Even the choice of instrument can give us clues as to what is going on.[7] If Doris Day is parking her car to the sound of a bluesy sax, she's about to meet her lover, but if the music is all soaring violins and piano, then she's meeting her husband. If she's parking to the jolly sound of banjos, her husband is a hillbilly, and accordions would, of course, mean that she's parking in Paris. (If she's just about to marry her hillbilly lover in Paris, then the orchestration is going to get a little complicated.)

Many of these clues are culture-specific. Brass instruments in Hollywood films often indicate bravery, but in a Hindi film they mean that villainy is afoot.[8] And if you're not Indian, you'll probably have missed out on the fact that the title music for *Monsoon Wedding* is based on the traditional Indian music for the groom's procession at a marriage ceremony.

In the UK TV series *Doctor Who,* the main character is a highly intelligent time-traveling hero. Since the program first came out, the writers have had to rely on the fact that the Doctor has an assistant who is quite often baffled by what's going on at any particular moment. The assistant has three main jobs as far as the plot is concerned: (1) to be rescued by the Doctor, (2) to rescue the Doctor, and most important, (3) to ask the Doctor dumb questions about what's going on so that the TV audience can benefit from the information.

Assistant: What's happening, Doctor?
Doctor: If I can't get this Hyperdrive transgrigulator negatively confabulated in the next forty seconds the whole galaxy will turn to ice.

The only real message here is "What I'm doing is URGENT!" The occasional implication of urgency is extremely important to most film plots, and the storyline can't always provide a spoken explanation or a view of a ticking bomb.

Music is an excellent way of indicating the level of urgency in a situation, and it's more enjoyable than over-explanatory dialogue. Composers use driving rhythms and variations in loudness to keep us on the edge of our seats—before delivering the relaxing chords after the crisis has passed. One method of increasing the tension in a film is to gradually increase the loudness of the music.[9] If we hear any noise getting louder, we automatically go on the alert, because an increase in volume generally means that something is getting nearer. As a survival reflex, our brain subconsciously assumes a worst-case scenario ("Something is going to attack!") and organizes the release of adrenaline so we can deal with the danger. As I mentioned in chapter four, this response can't be switched off, even if the increasing loudness is just a movie sound track and not a stampede of elephants. Pumping up the volume also creates the impression that things on the screen are happening faster, though interestingly this effect doesn't work in reverse: music that gets gradually quieter does not slow the action down.[10]

An important point here is that it is much easier to distance yourself from a visual image than from a sound.[11] Music helps us to forget about the screen and become immersed in the film. If you want to make a moviegoer jump with fear, it is far easier to do so with a sudden sound (musical or otherwise) than a sudden image—and of course the two together work best. If you'd like to test how effective movie music is in keeping you on the edge of your seat, just try watching a creepy (no dialogue) scene from any horror movie with the sound turned off. (The 1922 vampire movie *Nosferatu* is pretty frightening with the spooky music and pretty comical without it.)

One of the main jobs of movie music is to give certain passages continuity or, alternatively, to divide a film up into segments. Films often present us with a sequence of images that are not obviously connected, and music can help to glue everything together. For example, a sequence of different views of what the main characters are up to at lunchtime in various parts of a city will be held together by the regular pulse of the music. At the same time, the music will smooth over the fact that each of these short scenes would normally need to have completely different background sounds (one person driving in traffic, another in a busy restaurant, a third walking through a quiet park).

Flashbacks are another case in which music can hold everything together, but it can also be useful in separating the flashback from the main time frame of the film. You can even use the music to identify the era of the flashback. The main thing here is to play music from the appropriate decade, but directors can also degrade the quality of the sound to make it feel older—replicating, for example, the tinny quality of a 1930s record player. Another common application of music is as a background during slow-motion sequences. In this case the director requires music because the alternatives would be silence, or slowed-down sounds and speech—which is effective in only a very limited number of cases, such as

"D...o...n'...t t...o...u...c...h t...h...a...t s...w...i...t...c...h."

One way psychologists can tell how engaged you were in a film clip is simply to ask you how much of it you remember. A team of researchers led by psychologist Marilyn Boltz studied people's response to various film clips in which a sad or happy event was either accompanied or preceded by happy or sad music, and then checked to see how well the participants remembered each clip. The results were quite clear: if music accompanied the action, the "correct mood" music (e.g., happy music with a happy event) helped people become engaged with the film clip and remember it in more detail. Surprisingly, if the music was heard only in the seconds leading up to the event, the "wrong" music was more effective: a happy event seems more moving if it's preceded by worrying or sad music, and the emotional effect is even stronger if things turn out badly after we've had our hopes built up by happy music. This result clearly demonstrates that we use music in film to build up expectations—and if those expectations turn out to be wrong, we often experience a bigger emotional effect than if our forecast had been correct.[12] Surprise amplifies the emotion involved.

Movies take us on emotional journeys, so much so that the psychologist E. S. Tan calls film "an emotion machine."[13] Unlike moods, emotions require a trigger, and films give us plenty of emotional triggers: joy at the reunion of a mother and child, fear during a chase, relief after an escape. One way of looking at the combination of film and music is that the film provides a sequence of events to get emotional about, and the music deepens your emotional experience. Even when the action on the screen has no emotional content, music can be used to trigger an emotional response. One study scanned the brains of a group of viewers as they watched an emotionally neutral film clip and found that if fearful or joyful music accompanied the clip, the appropriate emotion centers of the brain were triggered, but there was a minimal emotional response to the film without the music, or to the music without the film.[14] This is very useful to filmmakers, since life (and therefore any film) is full of periods when the visual action is emotionally neutral. By using the right type of music, the director can tip us off that this walk to the shops is going to end tragically or romantically.

Film composers sometimes use a trick that was a favorite of the opera composer Wagner. The idea is that the major characters in the story have their own little signature tune, called a leitmotif. So, for example, in the 1938 film *The Adventures of Robin Hood,* King Richard has a rather regal-sounding tune, Robin's tune is brave and dashing, and Maid Marian's is pretty and innocent-sounding. The baddy—Guy of Gisbourne—is, as you might have guessed by now, accompanied by a menacing, dissonant theme.[15] In the heyday of Hollywood the composer was often given as little as three weeks to produce a big orchestral score, so leitmotifs became popular among composers because they automatically give structure to the sound track. Having written a collection of leitmotifs, the composer can then change the harmony, pace, and orchestration of the tunes to match the mood and rhythm of each situation. Leitmotifs are still used in some films because they help to bring coherence to a wide-ranging plot. You're probably familiar, for example, with the pompous military tune that accompanies the Empire forces in the *Star Wars* films, and leitmotifs played an important role in the music for the *Lord of the Rings* trilogy, from the gentle Gaelic-influenced Shire theme to the drama of the Rohan tune.

Whether or not the composer uses the leitmotif technique, there will be one or two main themes used throughout the film, and these will need to be adjusted to fit various moods as the story unfolds. Even in a remorselessly upbeat film called something like *Timmy Has a Lovely Time,* there will be moments of unhappiness or tension as we watch little Timmy burying his goldfish, or getting arrested for selling drugs at school. At this point in the action the happy main theme will need to be altered to make it a little less smug.

One of the commonest methods of turning a happy tune into an unhappy or tense one is to rewrite it in a minor key. This involves adjusting the jumps between certain notes—but the tune is still generally recognizable because it has the same rhythm and almost the same contour of up-and-down movement as we move from one note to the next. This technique was used to good effect in the 1942 film *Casablanca*. The composer, Max Steiner, incorporates the French and German national anthems into the score, and, like most national anthems, both of these tunes were originally written in a major key. But in order to reinforce our negative view of the Nazis, Steiner transposed the German national anthem into a minor key and provided a discordant accompaniment, giving the tune a menacing, oppressive feel.[16]

In some movies the characters in the film can also hear the music that we, the audience, can hear. The actors might be listening to their car radio or attending a concert, and in some cases this "in film" music (the technical term is *diegetic* music) can be used as a springboard for ordinary movie music. For example, a pianist playing in the film might stand to dance with his girlfriend and the music keeps playing, now with an orchestral accompaniment.

One of the cleverest uses of "in film" music happens in Hitchcock's 1956 film *The Man Who Knew Too Much*.[17] Our hero, Jimmy Stewart, knows that there will be an assassination attempt on a statesman during a concert. At the beginning of the climactic scene, we see the concert begin and watch the orchestra playing Arthur Benjamin's *Storm Clouds* Cantata. Cut to loveable, brave Jimmy as he rushes into the concert hall to look for the would-be assassin, with the music still playing in the background. There is no dialogue during this portion of the film, and besides matching the hero's anxiety, the music is synchronized with the action, calming a little as the hero meets his wife, and taking off again when he leaves her to continue his hunt for the bad guy. A minute later the assassin draws his gun and at the same time the percussionist picks up his cymbals, and from then on you know that the gunman intends to use the cymbal clash to cover the noise of the shot. It's particularly gripping when the percussionist prepares for his big moment by raising the cymbals to shoulder height and . . . but I won't spoil it all by telling you what happens next.

We usually experience "in film" music differently from normal movie music. In our daily life our brain is forever trying to work out what's going on around us, and one way we do this is by combining what we are seeing and what we are hearing together into a single stream of information. If we are walking into town, the sight and sound of a violin player on the street become fused into a single thought: "There's a busker over there. What a happy tune." If we are watching a film with "in film" music and we see and hear a busker playing a similar tune, we will come up with exactly the same thought. But if we are watching a film with a normal sound track that consists of that same happy violin music played over a clip of our hero walking down an empty street, we think, "Our hero's in a good mood." Because we are unable to link the music to a "real" source in the film, we focus on its emotional content, which we then superimpose onto the visual action.

Film directors have differing levels of interest in the music side of their films. Some directors—including Charlie Chaplin and Clint Eastwood—have even composed music for their own films. It's quite common for directors to team up with a favorite composer, as Hitchcock did with Bernard Herrmann, the composer who, taking his inspiration from the sound of human screaming, wrote the famously terrifying "eee! eee! eee!" music for the shower scene in *Psycho*. Although the two men usually worked very well together, they completely disagreed about what was needed for this scene. According to Herrmann, Hitchcock's instructions to him were "Well, do what you like, but one thing I ask of you: please write *nothing* for the murder in the shower."[18] Luckily for us Herrmann ignored the maestro, and a masterpiece was born. Since then composers have often conveyed terror in much the same way—with no melody, jerky rhythms, and high, dissonant harmonies. In fact, the shower scene music has become so iconic in its own right that it's used as a standard terror signal.[19] For example, the animated Pixar film *Finding Nemo* uses the shower scene music as a leitmotif for Darla—the little girl that all the fish are (justifiably) terrified of.

We can be thankful that, in this case, Hitchcock eventually agreed that Herrmann's ideas about music for the scene were better than his own, but things don't always work out so well for the composer. Movie music is usually completed after the filming ends, which means that it's not available when the director is editing the movie. In order for the film to sound more like a finished product, and to help the composer get a feel for the type of music the director wants, the director often records a "temp track" of suitable music to accompany various parts of the movie. The temp track is frequently a collection of classical pieces that match the moods of the various scenes. Working day after day on the film edit, the director can get used to the temp track and, in some cases, will prefer it to the newly composed stuff written especially for the film. Two famous cases of this are the films *2001: A Space Odyssey* and *The Graduate*. The director of *The Graduate,* Mike Nichols, was a big Simon and Garfunkel fan and wanted them to write a collection of new songs for the film, but in the meantime he made a temp track of existing songs. Ultimately the sound track combined some of the new stuff ("Mrs. Robinson") with some of the original temp track ("The Sound of Silence"). In the case of *2001: A Space Odyssey,* Stanley Kubrick decided to use his temp track of classical pieces for the finished film. This must have been extremely irritating for Alex North, the composer who had written and recorded a complete original sound track—especially as he only found out that his score had been dropped when he attended the opening night of the film! Kubrick also managed to irritate another composer, György Ligeti, by using two of his pieces in the film without getting his permission or giving him any acknowledgment.[20]

We humans are extremely prone to making up stories about whatever we see going on around us. Even when we are watching leaves floating down a stream, we sometimes look for types of behavior and might use human motives (e.g., competitiveness) and personality stereotypes (e.g., confident or shy) to fit a story to the situation. The fact that we ascribe motives to inanimate objects was proved fairly conclusively by a silent film that psychologists Fritz Heider and Marianne Simmel made in 1944. The film, which is only one and a half minutes long, follows the adventures of two triangles and a circle as they move in and around a hollow square. Ever since the film was made, viewers have been automatically giving the three shapes personalities and motives and building up stories about what's going on. The big triangle is widely regarded as a bully, and the other two shapes attract varying amounts of sympathy. You can watch the film on YouTube if you search for "Heider and Simmel." I just did so, and I can confirm that the big triangle is indeed a bad-tempered swine who deserves a good thrashing. Mind you, that small triangle was asking for trouble.

Years after the film was made, another pair of psychologists, Sandra Marshall and Annabel Cohen, were looking for short movies for their experiments into the effect of music on our perception of what's going on in a film. The Heider and Simmel film seemed like a good choice. In 1988 they played the film to three separate audiences: one group watched the film in silence, and for the others it was accompanied by either "weak" or "strong" music.* The

* Sandra Marshall composed the music for the experiment. The "weak" music used a major key at a moderate tempo and generally used only one note at a time. The "strong" music was in a minor key with a slow but accelerating tempo and more than one note at a time.

psychologists initially thought that music would affect the viewer's attitude toward the film in general—that strong music might make all the characters in the film seem more active, or more evil, while the weak music might have the opposite effect. The actual results were therefore a little surprising.

When the audience watched the film in silence, or with the weak music playing, they perceived the big triangle as being more active than the small one. When the strong music was played, however, the small triangle came across as the more active of the two (although it was still being bullied by the bigger one). The music was having different effects on the different "characters" rather than affecting the film as a whole.[21] Marshall and Cohen looked into what was happening and came up with an intriguing theory that has gained a lot of influence in the intervening years.* It seems that our brain is involved in a two-stage process. First of all, our attention is drawn to an object whose movement has something in common (or nearly in common) with the rhythm of the music playing. If the music is fast, we focus our attention on whatever object is moving around quickly. Once we are focused on the object in question, our brain projects onto it the characteristics of the music—happy, fearful, sad, etc.[22] In this case, the rhythm of the strong music fit best with the actions of the small triangle, so it came across as being very active. The actions of the large triangle didn't fit with the strong music, so it didn't grab the attention of the viewers, and the fact that it actually does move more than the small triangle was ignored.

Film directors and composers often use this trick of drawing attention to something or someone on the screen by making some

---

* The Congruence-Association Model, or CAM. For Professor Cohen's full description of the model, read chapter two of *The Psychology of Music in Multimedia,* ed. Siu-Lan Tan, Annabel Cohen, Scott Lipscombe, and Roger Kendall (Oxford University Press, 2013).

aspect of the music synch with the movement of the object or person in question. This can be done purely rhythmically—as when Rocky (in *Rocky II*) runs to the beat of the music through the streets of Philadelphia and up the art museum steps.[23] The on-screen action can also be mirrored in the contour of the melody, as in the rowing scene in *Ben-Hur,* where the melody follows the up-and-down movement of the oars.[24] Rocky isn't always exactly in time with the music, but we are quite forgiving of minor flaws in synchronization. An action and its accompanying sound can be out of synch by nearly a quarter of a second* before we find it noticeably peculiar.[25]

This is probably because we're used to some lag time between what we see and what we hear. In real life, if we see a man one hundred yards away hit a fencepost with a hammer, the sound reaches our ears about a quarter of a second after the visual information reaches our eyes, because sound travels much slower than light. On top of that, we don't process aural and visual information at the same rate. Audible and visual information go through different biochemical processes before they get transformed into electrical messages that whizz off to the brain. Our hearing system is faster at passing information to the brain than our vision system: visual signals take about a twentieth of a second longer to process than sounds.[26]

So, taking into account the different speeds of the sound and light as they travel toward your head, and the different processing speeds for the two types of information, the only time your brain gets exactly simultaneous eye and ear information is if the bloke hitting the fencepost is about fifty feet away.[27] At every other distance the brain has to ignore the fact that there is a slight difference in timing between the sound event and the visual action. The

---

* If you're wondering how long that is, it takes about a quarter of a second to say the word "it."

sound information arrives at our brain a little late if the action is farther than fifty feet away, and a little early if the action is nearer than that. So the fact that sounds and actions don't quite match up in films doesn't bother us too much because we have an automatic tendency to mentally synchronize what we see and what we hear.

Which brings us to something very peculiar—and an experiment you can all do at home.

Put on a film, any film, and turn the sound off.

Now pick some music, any music, and play it as the sound track to the film.

In the words of psychologist Herbert Zettl, "you will be amazed how often the video and audio seem to match."[28] This is because there is generally a lot going on in any film and we tend to focus on actions that coincide with the beats, phrases, and climaxes of the music we're hearing.

Obviously some combinations will disappoint even the most hopeful watcher, but you'll be surprised how often things seem to synchronize very well. Oddly, you don't need to start the music at any exact point in the film. Actions will appear to become synchronized quickly wherever you start the music. But how can that work?

Say you are watching a film of a tennis match to the sound of a random pop song. The music may, after a few seconds, synch with some of the foot movements of the players. After several seconds the music–foot movement synchronization will be lost, but now the music is in time with the impact of the ball and the racquet, and a few seconds later the beat fits the timing of the ball hitting the ground... Spooky...

Actually it isn't spooky at all. *Some* feature of the action is likely to be approximately in time with the music at any point. Our brain likes things to be simple, and it requires less mental effort to combine a certain visual action with a sound rather than having to process them separately.[29]

If you want to see just how uncannily a random sound track can synch with a film, try taking the advice of Professor Scott Lipscomb from the University of Minnesota.[30] Put on the 1939 film *The Wizard of Oz* and, at the beginning of the MGM lion's third roar, press "play" on Pink Floyd's 1973 album *Dark Side of the Moon*. Now sit back to be enthralled.

CHAPTER 8

Are You Musically Talented?

A lot of people would like to take their love of music one step further by learning how to play an instrument, but they are put off by the idea that they don't have the required talent. This is a shame because, as this chapter will show, this "I haven't got the talent" worry is irrelevant. The strange truth is that most professional musicians haven't got any innate musical talent either.

The word "talent" can be used in several ways, but the two most common reasons people use it are:

1. Wholesome pride in someone else's skill. ("Yes, my wife and I are very proud of little Jessica's talent for seal hunting. Here's the club we gave her for her eighth birthday.")
2. A peculiar combination of pride, laziness, and the bitter resentment at the unfairness of life that we all feel from time to time. ("No, I wasn't picked for the team. My idle brother was—but he just has a natural talent for the game.")

In both cases the talent is assumed to be a gift that people are born with. They deserve a little bit of credit for it, but not too much, because it's just the luck of the draw. Some people have curly hair, and some people have a talent for ice sculpture.

Music is a field in which the word "talent" is bandied about a lot: the world is full of "talented" violinists, conductors, and rock guitarists. Obviously no one is born with the ability to play the violin; like everyone else, a talented person must learn the instrument. But the general view is that those with talent will learn much faster and more easily and become far more proficient than the pitiable untalented folks could ever be.

Well—I have good news and bad news.

The good news: talent is mostly myth. So now you can take even more pride in your heroes and children, as they probably weren't born with extra skills.

The bad news: talent is mostly myth. So now you no longer have an "I'm not musically talented" excuse not to start those piano lessons you've always thought would be pointless in your case.

But what on earth can I mean? Some people are clearly better at music than others. So if they're not talented, what are they?

Let me tell you a story.

In 1992 a team of researchers in the UK decided to do some serious research into musical talent. Professor John Sloboda and his team investigated 257 young musicians who ranged in ability from those who had studied an instrument for only a few months and had given up, to those who were actively training to be professionals.[1]

Fortunately for the researchers, they had access to an accurate measure of the abilities of the individuals involved—the UK grade system. If you take music lessons in the UK, you are encouraged to take grade exams every year or so until you reach the top grade, grade 8. At this level you are pretty good at your instrument and capable of giving a concert or playing at your sister's wedding without engendering cringing embarrassment all round. This is also the grade that is a requirement for studying at most music colleges.

So the researchers knew the musical education performance history of 257 young musicians of different levels of ability, and they knew when they had passed their various grade exams, so they could compare how good they all were.

The subjects of the study were divided into five groups:

The top group of musicians had gained entry to a high-level music college by taking part in a competition. These people were training to be professional musicians. We'll call them the "A group."

The B group students were good musicians but they hadn't done well enough in the competition to get into the college.

The C group students were serious about music and had thought about applying to the college, but eventually decided against entering the competition.

The D group students were learning a musical instrument for fun but weren't considered (by themselves or anyone else) to be music college material.

The E group students had started learning an instrument but had given up.

It sounds pretty obvious that the A group of students, who succeeded in the competition and ended up training to be professionals, would, on average, be more talented than the B group, who would be more talented than the C group, and so on. So Professor Sloboda and his colleagues donned their computers and booted up their lab coats to look at how quickly the talented students rose through the grades compared to their less gifted compatriots.

When they looked at the figures and interviewed the students and their parents, they found what they expected: the high achievers *did* get through their grades faster than the others. After three and a half years of training, the A group had, on average, achieved grade 3, whereas, in the same amount of time, the C group had only achieved grade 2.

But when the academics looked a little closer, they began to suspect that talent was *not* the key to success. The numbers showed that, on average, the group A musicians needed almost exactly the same number of hours of practice as any of the other groups in order to pass the next grade exam. The average amount of practice any of the students had to put in to get from grade 1 to grade 2 was two hundred hours, no matter what group they were in. To get from grade 6 to grade 7 took on average eight hundred hours of practice. The average total amount of practice needed for *any* of the students to go from total beginner to achieving grade 8 was just over three thousand hours (although, of course, they didn't all go that far).

The conclusion was simple: the more you practice, the faster you become a good musician. The only "gift" the A group students had was the gift of diligence: they started off practicing more than the other groups and also increased the amount they practiced as the years progressed. This group started off doing about half an hour's practice a day in their first year of learning their instrument, and increased to over an hour a day by their fourth year. The lower-achieving students started off doing less than half an hour and didn't increase the amount of practice time much in subsequent years. (For example, group D started at only fifteen minutes a day and rose to the dizzying heights of twenty minutes over those initial four years.)

On average, group A students weren't especially talented; they just put in more hours of work every week.

The results of the Sloboda study were confirmed by another group of psychologists, who carried out a study of music students in Berlin in the early 1990s.[2] The researchers began the project by asking the staff of the Music Academy of West Berlin to rank their violin students into three groups—let's call them excellent, good, and ordinary. The researchers then analyzed how all the students spent their time on an hour-by-hour basis and also looked into the history of their musical training. They found that the students were very similar in many ways. They had all started their training at about the age of eight, and they all spent about fifty hours a week involved in various musical activities.

The only big difference between the groups was how much solo practice they did. The excellent students had, on average, 7,410 hours of solo practice under their belts by the time they were eighteen years old, compared to 5,301 hours for the good students and only 3,420 hours for the ordinary ones. These figures fit in well with the generally accepted rule that just about anybody can achieve a professional standard in nearly any skilled activity—from athletics to zoology—if they put about 10,000 hours of practice into it (and in case you're wondering, 10,000 hours is equivalent to about four hours a day, every day, for seven years).

For those of you who were keen on the concept of talent and resent the idea that musical achievement can largely be attributed to simple, boring hard work, please don't forget, this makes the high achievers *more* admirable—not less.

When parents proudly describe the musical talent and potential of their beloved offspring, they are, without realizing it, actually talking about how well the child has *already progressed* on the instrument in question.[3] They don't point at little Henry before he's laid his sticky fingers on a violin and say, "He looks like he would be a marvelous violinist." They actually wait until the child has acquired some skills and then declare his genius for playing "Mary Had a Little Lamb" or "Smoke on the Water." They seem to have forgotten the weeks of squeaks and all the hard work involved.

The key to acquiring high-level musical skills is something called *deliberate practice.* The more deliberate practice you do, the better you get—and this applies to any skillful activity. But deliberate practice is not the same thing as ordinary practice. Ordinary practice often involves simply repeating something you can already do pretty well. Deliberate practice, by contrast, means that you are taking a step forward. You are doing something you find difficult—and once you have mastered it, you will be a step nearer to perfecting your skill. One of the defining characteristics of deliberate practice is that generally it isn't fun—which is why excellence is rare.

The film producer Sam Goldwyn once famously said, "The harder I work, the luckier I get,"* and for musicians this could be reworded as "The harder I work, the more talented I get."

But that's not the whole story.

You may have noticed that I've only been referring to the average behavior of the various groups so far. But there were some students who were far from average for their group. Some students spent a lot more time practicing than the average for their group, and some were successful even though they practiced far less. One of the most interesting results of this survey was that *in every group* there were "a handful of individuals"[4] who were successfully passing their grade exams even though they were putting in *less than a fifth* of the average hours of practice within their group. If talent really exists, it must be among this lot.

But what is the nature of their talent? Are they talented at music? Or are they talented at practicing effectively? And why, if they are so talented, aren't they all in the A group?

We might be picky and say that they are just better at practicing. Maybe in every one hundred hours of practice they manage to cram in ninety hours of deliberate practice, whereas the others achieve only thirty hours. Maybe they enjoy the challenge of deliberate practice—the tedium of repeating a difficult passage until it's fast and smooth. But in that case, wouldn't this just be a different definition of talent? The usual definition of the word is

---

* Goldwyn was reputed to be a rich source of one-liners, including "Include me out" and "A verbal contract isn't worth the paper it's written on." But the work-luck line isn't completely original. It's a paraphrase of a quote from Thomas Jefferson, who said, "I'm a great believer in luck, and I find the harder I work, the more I have of it."

"natural ability," and this could easily apply to a natural ability to do the work involved.

John Sloboda and his team also looked into the influence of the first music teacher the students had as children.[5] They found that one of the most important factors in the eventual achievement of professional-level playing was how much fun a student's first teacher was, and how friendly. Later on in their training the fun factor was less important than how skilled the teacher was, but it's fairly clear that in the early stages, kids will work harder to please a teacher they like. The reverse is also true: I have met lots of people who gave up musical instruments at an early age because they were taught by people who would have been more at home working for the secret police of a particularly insensitive fascist state.

In 2003 psychologists Susan Hallam and Vanessa Prince asked over a hundred professional musicians to complete the sentence "Musical ability is..." and found that the vast majority of the replies they received centered on words like "learned" and "developed." Non-musicians, answering the same question, tended to finish the sentence with phrases implying that there was some sort of natural gift involved.[6]

However we define it, there obviously are some people who are musically talented—but they are few and far between, even among professional musicians. John Sloboda's results imply that only about ten members out of the hundred or so who make up a symphony orchestra could truly be considered talented. In most cases this simply means that they achieved the same (very high) level of playing as their colleagues with far less practice. In a very few cases, however, they put in the same number of hours as their peers and became solo instrumental stars.

The proud parents of the other 90 percent of highly skilled professional musicians probably described them as talented, but this description is wrong. It would be far more accurate simply to give them the credit they deserve for all the work they put in.

At the other end of the spectrum from people misusing the word "talent" are those who inaccurately declare themselves "tone-deaf." Tone deafness is a problem—or at least a musical disadvantage—for those who really have it, but thankfully it's quite rare.

Tone deafness

Tone deafness—more formally known as amusia—is an inability to perceive or reproduce melodies. You can either be born with amusia or you can become "amusic" as a result of brain damage. According to the Montreal Battery of Evaluation of Amusia test, which checks your perception of the various components of melody (rhythm, key, contour, etc.), only two or three people out of every one hundred are amusic.

In Western music the smallest jump in pitch in a tune is a semitone, the distance between two adjacent notes on a piano. If two notes a semitone apart are played one after the other, most of us find it easy to tell which one of the two is higher. In fact, most infants can judge between two notes that are only half a semitone apart.[7]

It's possible to make the jump in pitch between two notes much smaller than a semitone by tuning two strings on a harp until they are almost the same pitch. If we did this, starting with a semitone and making the pitch difference smaller and smaller, there would come a point at which a normal listener couldn't tell which of the two notes was the higher one. Eventually you could make the jump so small that it would baffle even an expert listener. As long as the jumps between pitches in the melodies we hear are a semitone or more (and they always are), we can spot the differences in pitch between the notes and follow the ups and downs of a tune. We can then remember the tune and sing it if we want to. If the jumps in pitch were much smaller, though, we would run into trouble.

For example, let's imagine that we have tuned a standard forty-seven-string orchestral harp so that each string is much less than a semitone higher in pitch than the one next to it. In this situation, harp string number two is only a tiny bit higher in pitch than string one, and string three is a tiny bit higher than string two, etc. If someone plays us a "tune" on this thing, we will have two problems. If the tune involves plucking adjacent or nearby strings, we won't be able to follow whether the pitch is going up or down between notes. Also, if there is a big jump in the tune from, say, string four to string thirty-five, we'll hear that the second note is higher, but we won't be able to tell if it involves string thirty-five or thirty-six or thirty-four; we'll just know that it's somewhere in the mid-thirties. So if music used really small pitch jumps, half the time we wouldn't know what was going on at all, and the rest of the time we would have only a vague idea.

This is what standard music is like if you have amusia. The problem for amusics is that even a semitone is too small a jump for them. They find jumps of around a semitone so small that the notes can't be clearly distinguished. Bigger jumps are more obvious but can only be estimated inaccurately. So if someone with amusia tries to sing a song they've just heard, they are flying blind. Sometimes they don't know whether the tune goes up or down next, and if they do know which way it goes, they can make only a rough estimate of how big the jump in pitch is.

How talented are you?

The unfortunate two or three people in every hundred who have amusia obviously have a serious disadvantage when it comes to listening to, or trying to create, music—so they could be considered musically untalented. At the other end of the scale we find an equally small number of lucky buggers who acquire musical skills quickly and easily. Whether or not this is because they are simply more efficient at practicing, we don't know, but we do know that the ability is not always linked to eventual high achievement.

We also know that the vast majority of professional musicians had to put in thousands of hours of practice to acquire the skills involved.

Whether or not you are a musician at the moment, the statistics speak for themselves. Musicianship is almost certainly within your grasp if you start teaching yourself or taking lessons—but make sure you get a teacher you like, and that you're having fun in the lessons. If you're going to teach yourself, make sure you enjoy the book or website you're using. The early months are quite tough because it's all new, so everything you do falls under the definition of deliberate practice (which, as we know, is not fun). But after the initial trials and tribulations, it all becomes much more enjoyable.

I've mentioned that it generally takes about ten thousand hours of practice to become a professional musician, but if you have no intention of performing at Carnegie Hall, you can get to the point where you really enjoy playing in just a few weeks. If you make it your New Year's resolution to do just a couple of hours' music practice a week, you'll be able to play a couple of pieces for your friends by springtime.

Those of you who are interested in making a start and have access to a keyboard or piano might enjoy a video I made on YouTube, rather bravely titled "Learning to improvise at the piano in an afternoon—for people who can't play the piano, can't read music, and don't know how to improvise."

If you have any feedback for me at the end of the afternoon in question, please feel free to email me at howmusicworks@yahoo.co.uk and I will try to answer any questions you might have.

CHAPTER 9

A Few Notes about Notes

We draw the emotional content of music from melodies and harmonies that are, of course, sequences and combinations of notes. And to continue our discussion of music, we now need to know a little more about notes themselves. This short chapter will explain the basics of what a note is and—most important—the whys and wherefores of the octave. Once we've covered this ground, we can move on to discuss the ins and outs of how our brain responds to melodies and harmonies and why we have a natural aversion to dissonance.

First of all, what is a note?

Your eardrums are like tiny, highly sensitive trampolines that react quickly to changes in air pressure. Unless you are in a completely silent environment, the air near your ears is continuously flickering up and down in pressure hundreds or thousands of times every second. As long as these rippling pressure changes happen more frequently than twenty times a second and less frequently than twenty thousand times a second, we experience them as sound.* Whenever the pressure rises, your eardrum is pushed inwards, and when the pressure drops, the eardrum is flexed outwards a little.

When you hear non-musical sounds, like the scrunching up of a paper bag, there is no pattern to the pressure ripples involved. The

* Sound waves with pressure ripples below twenty times a second are known as subsonic. Vibrations above twenty thousand times a second are ultrasonic.

movement of the paper while being scrunched produces a very complex selection of pressure waves that flow across the room to your ear like ripples across the surface of a pond. When these pressure ripples reach your ear, they move the eardrum in an equally complex way. Eventually your brain compares this noise to its library of often heard noises and comes to the conclusion "Ah, something made of paper is undergoing a thorough scrunching somewhere nearby."

When you hear a musical note, however, there is a repeating pattern to the pushes and pulls on your eardrum. If you hit a piano key, twang a guitar string, or blow into a saxophone, you set up a trembling vibration in the instrument and the air around it. (You can feel this tremble if you touch any instrument when a note is playing.) The regular to-and-fro vibration of the instrument produces a pattern of repeating ripples of pressure in the air that push and pull on your eardrum. The pitch of the note you hear will be determined by how frequently the cycle of vibration takes place. For example, the lowest note on a piano will push your eardrum in and out fifty-five times every two seconds, and the highest note on a piano will do so at a rate of over four thousand times a second.

We get the *frequency* of a note from how frequently it pushes your eardrum in and out. If a note flexes your eardrum in and out in a repeating cycle 440 times a second, we say it has a frequency of 440 cycles per second. Generally the phrase "cycles per second" is

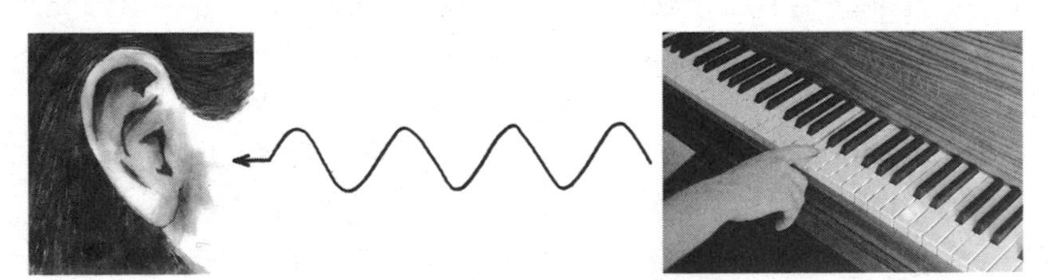

Ripples of pressure travel through the air from a vibrating musical instrument. When the ripples reach your ear, they push and pull on your eardrum in a rapid repeating pattern and you hear a musical note.

replaced by the word "hertz" in honor of the German scientist Heinrich Hertz. And "hertz" is usually written as the abbreviation "Hz." So the note that pushes your eardrum in and out 440 times a second has a frequency of 440 Hz.

In the previous image I've drawn a very smooth repeating pattern of up-and-down pressure, but the ripples from a real instrument are usually a lot lumpier than this. For example, here are typical repeating pressure patterns from a flute, an oboe, and a violin, all playing the same note.* These traces are taken from recordings of the three instruments, and the up-and-down fluctuations of the pressure they produce are described by the rising and falling of the traced line.

flute

oboe

violin

A musical note can have almost any pressure ripple pattern as long as that pattern repeats itself over and over again. As you can see, the repeating pattern for the flute is a fairly smooth double bump followed by a valley, whereas the oboe pattern fluctuates a lot more and involves two bumps followed by a double bump followed by a valley. This increase in complexity in the repeating

* The note "middle A" or "A_4"—which has a frequency of 440 vibrations per second.

pressure pattern is reflected in the different timbres of the two instruments. A flute note sounds pure and simple, and the oboe has a richer, more complicated timbre. All the notes I've drawn here have the same pitch because each pattern takes the same amount of time to repeat itself. (In each case I've drawn about four and a half cycles of the pattern.) For the note involved here, your eardrum would be flexing in and out to these complex but repeating cycles of movement 440 times a second.

While we're on the subject of timbre, it's worth bearing in mind that the timbre of musical instruments is not the same over their whole range. For example, a clarinet sounds rich and warm for low notes and much edgier for high notes. So the pressure patterns for any instrument are different shapes for different notes.

From now on I'll just use drawings of the simple up-and-down repeating pattern we saw first. But please bear in mind that real notes from actual instruments have far more complicated patterns.

When we listen to an unaccompanied melody, we hear notes of different pitch in a sequence. Higher-pitched notes have more pressure cycles per second.

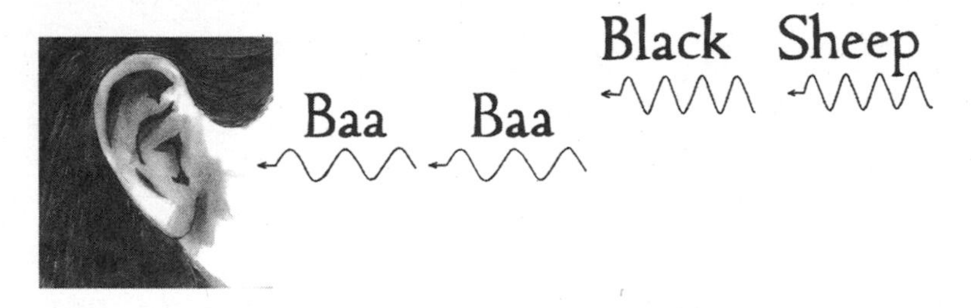

I'll be discussing melodies in the next chapter, but for the moment I want to talk about what happens when you play notes simultaneously. To make things as simple as possible, we'll just look at the situation in which only two notes are involved.

Obviously, if you play two different notes at the same time, one of them will have a quicker up-down pattern than the other, so the ear is kept busy trying to respond to them both simultaneously.

Depending on the notes involved, the resulting sound can be calm and relaxing or rather tense and anxious. Within any piece of music the composer or songwriter might choose an anxious combination of notes before moving on to a calm, relaxing combination, which is followed by more anxiety. This cycle of tension and release in the harmony is one of the tools musicians use to manipulate our emotions.

If you listen carefully to two notes played at the same time, you can generally distinguish the two individual notes involved. You could, for example, hum the lower one and then the higher one.

But—and it's a big but—something odd happens when we play any note together with the note an octave above it.

"And what," you might ask, "is an octave?"

At the heart of this question is a phenomenon that has been discovered by every musical society in the world and is the cornerstone of all musical systems throughout history. It's possibly the most important single piece of musical information there is.

It's called octave equivalence.

Octave equivalence

Musicians are told early in their training that notes an octave apart sound very similar, even though there is a big distance between them in pitch. (The first two notes of "Somewhere Over the Rainbow" are an octave apart.) But if, when you are learning an instrument, you ask *why* two such distant notes have such a strong resemblance, the response will generally be a single sentence along the lines of "They sound similar because the upper note has twice the frequency of the lower one."

Unless you are familiar with the concept of frequency, this answer is only marginally more informative than "That's just the way it is—OK?"

Octave equivalence is the reason why there are eight A's on a piano—and eight B's, and so on. If you look at a piano keyboard, you can see that there is a repeating pattern of keys from left to right, and the notes are named in a repeating pattern as well. The note names go from A to G sharp (often written G#), and then the letter names repeat. (We'll see why in a minute.)

To show which A or D or F we're talking about, each note is given a little number after the letter, going from zero for the lowest note to 8 for the highest. So, for example, A_4 has a higher pitch than C_3.

If you take any note as your starting point, the next note up (to the right on a piano keyboard) that has the same letter name is an octave above it: A_3 is an octave above A_2, and D_2 is an octave above D_1, and so on. (Carrying on in the same vein, E_4 is three octaves above E_1, and B_6 is four octaves above B_2, etc.)

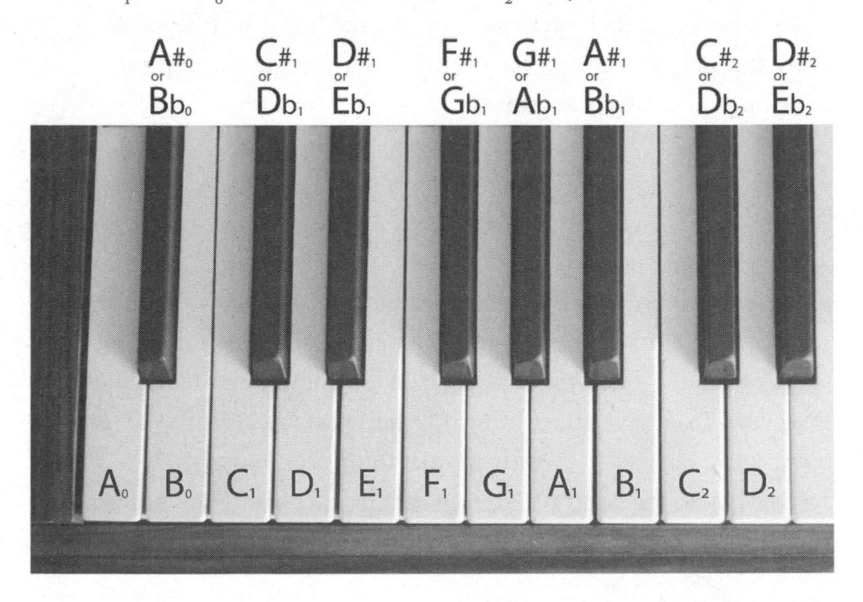

If you play two notes an octave apart one after the other, they sound oddly similar even though they are a long way apart in pitch. Also, if you play the lower of the two notes first and then add the higher "octave above" note to it, the two notes merge into each other seamlessly. If you try combining two notes that aren't an octave apart in the same way, they don't consume each other like this; they retain their own identities.

For those of you with memories as bad as mine, here is that drawing of the pressure ripples from a smoothly timbred musical instrument traveling into a listener's ear again. (I don't know about you but I find it irritating when authors ask you to flick back a few pages to find an illustration.)

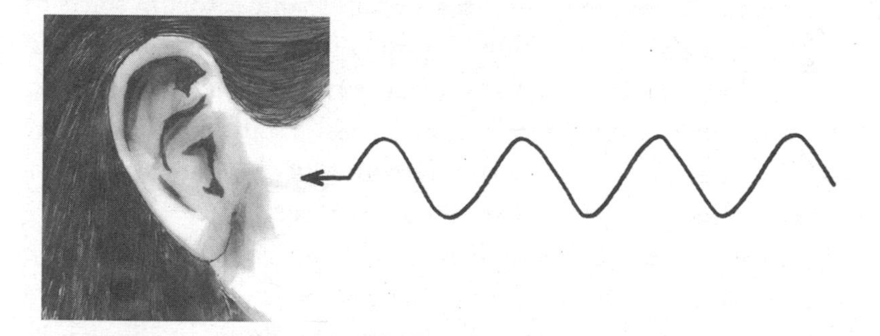

For our discussion of octave equivalence we are going to simplify things and look only at what happens during a single cycle of up-and-down pressure. Like this:

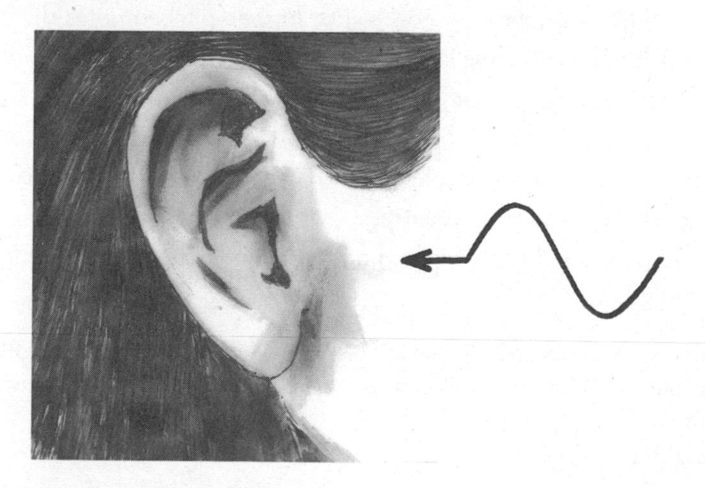

In real life you'd never experience this, but for our discussion a single cycle is useful because a real note is made up of a continuous stream of these cycles, so we need to discuss only one of them.

It takes a certain amount of time for our cycle to do its job on the eardrum. Let's assume that you are in a quiet room and this cycle is the first part of a note that is just about to arrive at your ear. At this point your eardrum is in its relaxed, flat state and the air pressure outside the ear is "normal." Now the pressure cycle arrives. Initially the pressure rises, so your eardrum is pushed inwards. After a short time the maximum pressure is reached and your eardrum starts to relax again as the pressure falls. But then the pressure continues to fall until it's below normal, so your eardrum flexes outwards. After the minimum pressure is reached, the pressure starts to rise to a normal level again and your eardrum returns to its flat state. This cycle repeats and repeats and you experience a musical note.

Let's introduce a bit of simple jargon here. The time needed for a complete single cycle of any note is called its *cycle time*. From a musical point of view, the most important thing about a note is its pitch, and the pitch of any note is directly related to its cycle time.

If the note is, for example, A_2 (the note you get from the second-thickest string of a guitar), the cycle time will be nine one thousandths of a second,* or nine milliseconds. For higher-pitched notes the cycle time will be shorter, while lower-pitched notes have longer cycle times.

If we now play the note an octave above this (A_3), we will find it has a cycle time of half that length: four and a half milliseconds.† It's this halving of the cycle time that gives two notes an octave apart such a special relationship.

If we play one note and then add another that has half the cycle

---

* Actually it's one second divided by the frequency of 110 Hz (9.0909 milliseconds), but we'll use the approximate value of nine milliseconds for this discussion.

† Actually 4.5454 milliseconds.

time of the first one, the timbre of what you hear will change—but not the cycle time of the experience. Two cycles of the higher note will fit nicely into one cycle of the lower one, so we still have an aural pattern that repeats at the cycle time of the slower cycle. And because the cycle time hasn't changed, the pitch hasn't changed either. Your ear doesn't need to worry about coping with two notes with different up-down agendas. They've become fused together into one.

One cycle of our original note:

plus two cycles of our "octave above" note:

have combined to create one cycle of a slightly more complicated new sound with the same cycle time, and therefore the same pitch, as the original note.

Let's look at several cycles together so we see the repeating pattern more clearly:

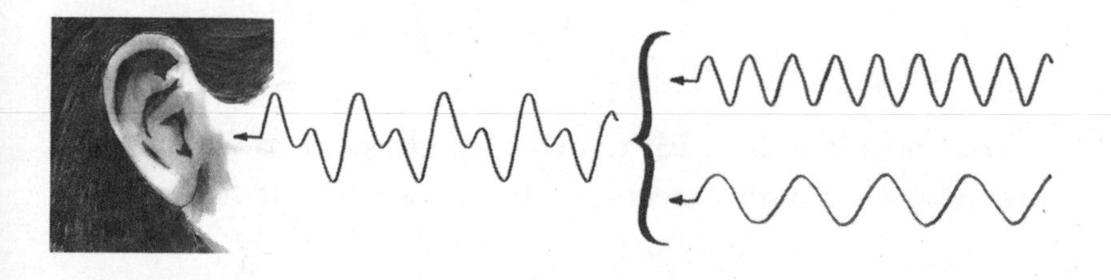

As far as our hearing system is concerned, the original note has swallowed the "octave above" note and merely become a more complicated personality in the process.

This phenomenon occurs with any two notes that are an octave apart. I happened to use A_2 and A_3 as my example, but I could have used D_6 and D_7 or $F\#_4$ and $F\#_5$, etc. Also, obviously, if it's true for A_2 and A_3, it's also true for A_1 and A_2, or A_3 and A_4, and so on. And this means that any A has a close family link with all the other A's, which is why we let them all share the same letter name. The link between notes that are several octaves apart (e.g., A_2 and A_6) is weaker than it is for a single octave, in the same way that your great-grandmother is less connected to you than your mother. But the family connection is still strong; the A's are a team, as are the B's, and so on.

The way notes an octave apart fit perfectly into each other is also the reason why they sound oddly similar if they are played one after the other. Imagine your ear has just heard a note with this sort of pressure pattern:

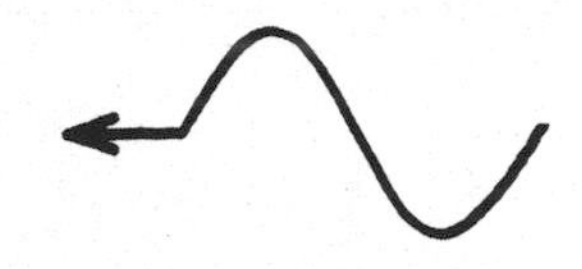

If the note an octave above is played next, your brain has to work out if this new note has half the cycle time or whether it has the same cycle time but with a double bump in its pressure cycle:

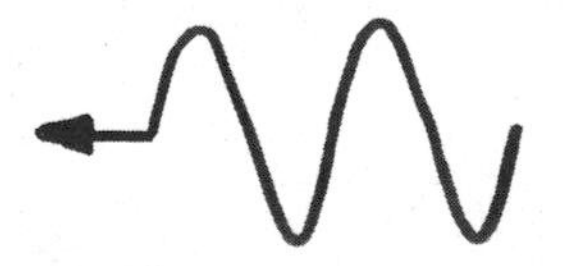

Your brain therefore has to make the choice as to whether this new note has a higher frequency or just a different timbre. This sort of confusion is possible only with notes an octave apart, and it's the reason why such notes sound weirdly similar.

OK, so notes an octave apart are a closely related group—but why do we use the word "equivalence"?

Let's say you are accompanying a singer on the piano. Your piano is in tune but some of the strings are broken, so one or two of the notes aren't working. The songwriter has written down that the first chord of the song should be G major and, as is usual, he has chosen exactly which notes he wants for the chord: G_4, B_4, and D_5.

But G_4 is one of your silent notes. "Quelle horreur!" you might pretentiously whisper to yourself as your fingers hover over the keyboard. Well, get a grip—all is not lost. G_3 or G_2 or G_5 will provide excellent replacements for your G_4. The chord will sound a little different, but the choice of notes required for that chord depends more on the letters than the numbers involved. So notes an octave apart are said to be equivalent as far as harmonies are concerned.

This equivalence, however, doesn't usually work for tunes in the way it does for chords. The replacement of a melody note with the note an octave above or below works on some occasions, but large hops like this tend to spoil the contour of the melody, and if you introduce several octave replacements in a tune, it rapidly becomes unrecognizable. Frequent octave hopping in the melody should generally be avoided unless you're a jazz soloist who's deliberately obscuring the tune in order to reveal it later (or as Kim would put it, "Unless you're a jazz soloist who's deliberately trying to make everyone's life a misery").

You may be wondering why, apart from the fact that notes an octave apart merge into each other and can sometimes replace each other, the octave is so important to our understanding of music.

Well, all musical systems around the world use notes that have different pitches. But how do we choose the sizes of the intervals, the jumps in pitch between the different notes? Wouldn't it be convenient if there were a naturally occurring interval that gave us a definite structure to build our system on?

You guessed it: the octave is a naturally occurring interval. By "naturally occurring" I mean that it can be produced easily from inanimate objects. If you take a bottle, hold it to your lips, and blow across its neck, you will get a pleasant low whistling sound—a musical note. Blow a bit harder and you will produce the note an octave above. If you are a prehistoric hunter twanging on the string of your bow, you will find that when you touch the vibrating string at its center point, it gives you the note an octave above its usual twang.

The octave is the basic building block of all musical systems, but it's a big interval; an untrained singing voice has a range of only about two octaves. So musical societies throughout history have devised ways of dividing the interval up into several smaller intervals. In the case of Western music, we eventually decided to divide the octave up into twelve equal steps and choose to use a selection, or key, of about seven of them at any one time (although we can change from one selection to another during the course of a piece of music). Classical Chinese music has opted for slightly different steps, with only five notes in any key.

But whatever the system involved, it's going to be based on the octave.

CHAPTER 10

What's in a Tune?

Musicians usually go to much better parties than music scholars, and as a natural result of this, there have always been resentful, power-hungry musicologists trying to impose their will on the musicians by inventing rules. They have, for example, tried to ban certain harmonies because they were "wrong." But these rules have always failed as new fashions take over from the old ones. For instance, back in 1890 it was considered wrong to fix your left hand in a particular chord shape and move it up and down the piano keyboard to produce what are known as parallel chords. What you were supposed to do was keep changing the type of chord you played as you moved your left hand from one chord to another. But the innovative young Parisian composer Claude Debussy liked the rather ethereal sound of parallel chords, and used them in several of his compositions. His professor's response is a great example of a bossy musicologist in action:

"I am not saying that what you do isn't beautiful, but it's theoretically absurd."

Thankfully Debussy wasn't put off. He simply replied:

"There is no theory. You have merely to listen. Pleasure is the law."[1]

At which point the professor probably gave him a C-minus for composition and Debussy huffed off to the nearest bar to write "Clair de Lune," one of the most beautiful pieces of piano music ever written—even if it is theoretically absurd by 1890 standards.

And Debussy was right: there are no hard-and-fast rules governing music; there are merely observations about what has worked in the past. Typically some scholar or other notices a pattern—for example, that pieces of music often end with a particular sequence of chords—and so they let it be known that "this sequence of chords creates a convincing ending." Meanwhile, musicians are using that pattern of chords without caring whether or not the musicologists are giving them the thumbs-up. This is how music theory has worked for the past several hundred years. The scholars and musicologists always follow in the footsteps of the musicians, simply analyzing what they have been up to. The music comes first, then the theory (except for one daft exception that I will be ranting on about later in this chapter).

So, the rules of how a standard melody is constructed are simply observations developed by looking at what happens in most tunes. Here are some of the more important ones[2] for Western music:

1. Most tunes are written in a major key.
2. Most tunes start on the keynote (e.g., B in B major).
3. Most tunes rise before falling back to the keynote again (although there will be some fluctuations within this overall "arch").
4. Tunes generally involve lots small jumps in pitch and very few big jumps. (More than half of the jumps will be to the note one step above or below in the key being used.[3] Repeating the note you are on—i.e., no step at all—is also fairly common. Jumps involving three, four, or five steps or more are progressively more unusual.)
5. A small step downwards in pitch is usually followed by another small step downwards.
6. If a big step is made in the tune, it will probably be upwards.
7. After a big step, the next step will probably be smaller and in the reverse direction.

8. There is a pecking order of how often the different notes in the key are used. For example, the keynote and fifth note in the key are used very frequently, but the note just below the keynote is used very rarely.
9. The keynote and the fifth note in the key are often used at points of rhythmic emphasis.
10. Notes that occur at a point where the melody changes direction (from rising to falling or vice versa) will often be emphasized rhythmically.
11. A melody will usually contain only one example of its highest note.

I have been trying all morning to write tunes that follow all of these rules at once—but it's almost impossible. Looking at my morning's output, I can tell you that as far as emotional content goes, they range from dull to very dull. Even the best one sounds like a cross between a not very popular nursery rhyme and an uninspiring hymn—a hymn for the patron saint of porridge manufacturers or something along those lines.

In practice, of course, most tunes follow only *most* of these rules. For example, "Baa, Baa, Black Sheep":

1. Is in a major key.
2. Starts on the keynote.
3. Follows an upward arch and descends back to the keynote.
4. Uses far more small steps than big ones.
5. Uses small steps downwards followed by more small steps downwards.
6. Has only one big step, which is upwards ("baa—black").
8. Follows the usual pecking order of note usage.
10. Uses the keynote and fifth note at points of rhythmic emphasis.
11. Has only one instance of its highest note.

But it doesn't obey rule 9. The change in direction in this tune happens on the first syllable of "any," which isn't particularly strongly emphasized in the rhythm. It also doesn't obey rule 7 about bouncing back a bit after a big jump.* There is a big jump right at the beginning of the tune—between the words "baa" and "black"—but instead of relaxing back to a note in the middle of this jump, the tune repeats the same note and then continues upwards. Naughty tune.

My attempts at rigid rule-following show that the rules do not automatically deliver beauty—it's just that a lot of beautiful stuff happens to follow most of these rules. We can use humor as an analogy to music here. You could put together a set of observations about the similarities among the acts of several comedians, and although the rules you develop might be true and useful, they won't necessarily help you to write something funny.

This is all pretty obvious when you think about it. If there were a formula for producing beautiful tunes, everyone would use it and all beautiful tunes would have a lot more in common than they do in reality. But this isn't to say that a tune formula hasn't been attempted; there are several computerized "melody generators" on the Internet. Unfortunately, the results are—to go back to the humor analogy—rather like hearing a "funny" speech written by a twelve-year-old who has read a book on writing funny speeches.

I'm not denigrating folks who write these melody generators. It's an extraordinary intellectual feat to get a computer to turn out even second-rate tunes. But the fact remains that they *are* second-rate—and, actually, I'm rather glad about that.

---

* If you want to know more about this "bouncing back" rule (including why it doesn't work in "Baa, Baa, Black Sheep") have a look at "Fiddly Details," section B, "Post-Skip Reversal."

Listening to a tune for the first time

When I say "we are all expert listeners," you might think I'm exaggerating, because from your point of view all you're doing is listening to the music and enjoying it. But if you want proof of your level of expertise, all you have to do is look at what's going on in your brain every time you listen to a piece of music you haven't heard before.

When you listen to a piece of music for the first time, you do so with a set of expectations. If you are familiar with the type of music involved (pop, 1940s jazz, reggae, etc.), you will know that certain note patterns and combinations are probable, and others are unlikely. Sometimes you will even be able to guess what the next two or three notes of the melody will be. There will be times, however, when your forecast of how the notes are going to progress turns out to be wrong, and you'll need to reassess the past few notes you've heard in order to continue making sense of the music. All this subconscious forecasting and checking is driven by our continual need to make sense of what's going on around us.

You might think this all sounds rather complicated. Isn't listening to music generally a passive activity? Don't we just follow the music and soak it up? Well, it depends on circumstances. If music is playing in the background when we're chatting with friends, we just let it wash over us, but what I'm talking about here is listening to a new piece by your favorite band or composer. In this case you are *listening* to the music, not just hearing it. Listening is a much more focused activity than hearing.

Let's assume you are a Western listener hearing (for the first time) a typical Western tune in a genre of music you are familiar with.

The first note starts. You don't know how long it's going to last or where it fits into the scale being used, but you'll already be making assumptions. You'll assume that it's part of a major key because most Western music is written in a major key. You'll also assume that the note is most likely to be the keynote, and if it's not, then it's likely to be either the fifth or third note in the scale.[4] Chances are you'll be right, because your assumptions are based on previous experience. You don't need to know consciously what a major key is or know anything about fifth and third notes in scales to make these assumptions. They are all made subconsciously.

Once the first note has finished, you will start making assumptions about the rhythm of the music. From now on, not only will you be making guesses about *which* notes are most likely to occur next, but also you'll be making assumptions about *when* they are likely to occur and how long they are going to last.

As far as rhythm goes, once the second note begins, you'll guess that it will be the same length as the first, or twice as long, or half as long, and you will often be correct, because that's how Western music is usually put together (although there will be occasional instances of notes that are one third as long, three times as long, or something else altogether). Western music is generally rather simple as far as rhythm goes, so these simple ratios (1, 2, and 3) cover most of it. You'll also assume that the overall pulse of the music will emphasize the first beat of groups of four or two—"ONE two three four, ONE two three four" or "ONE two, ONE two."[5] Once again this will usually be right, but if the music happens to be a waltz, this expectation will be wrong, because waltz rhythms are organized in groups of three—"ONE two three, ONE two three"—and if it's a traditional Irish jig, the overall pulse will repeat in groups of six or nine.

Except in the very simplest cases, we don't absorb an entire tune as a stream of notes; we divide it up into shorter phrases: "Baa, baa, black sheep" . . . "have you any wool?" . . . "Yes sir, yes sir" . . . "three bags full" . . . and so on. In some cases the notes are all the same length (as in "Baa, baa, black sheep" and "Yes sir, yes sir"), but phrases with different-length notes in them will usually end on a long note: "wool," "full."[6]

Our subconscious forecasts about phrases (e.g., "This phrase is coming to an end") and notes ("This will be the highest note in this phrase") are confirmed or contradicted only retrospectively, as the tune moves along. We can't even be sure how long any single note is until it stops. Sometimes we have to rapidly change our opinion of what we've just heard in order to make sense of a piece, in the same way that we instantly rethink the last few words of a conversation once we realize we've misunderstood something.[7]

Recognizing and remembering a tune from its contour

After we've heard something many times, we get a full appreciation of the rhythm and the notes involved, but we build up this accurate picture from an initial sketch or *contour,* which we memorize after just the first few hearings. The contour of a tune tells you the rhythm involved and whether the tune goes up or down in pitch at any point.

For example, the contour for "Baa, Baa, Black Sheep" can easily be seen in the written music:

Baa baa black sheep have you an - y wool? Yes sir, yes sir, three bags full.

There's a big jump from "baa" to "black," then the tune rises in single steps up to "an-...." From "an-..." there is a moderate-sized jump down to "-y," and then the melody comes down in single steps to the note it started on. When you first hear a new tune, your brain doesn't pay much attention to the size of the pitch steps; it just concentrates on whether the steps are upwards or downwards, and on the timing of the notes—the rhythm.

The importance of rhythm

Contours are the first step to remembering a tune—but they are also the main route to recognizing them. Obviously, if someone hums a tune with the proper rhythm and the right-sized up-and-down jumps, anyone familiar with the melody will recognize it immediately. But the odd thing is that the rhythm is more important than the size of the pitch steps. If you keep the rhythm correct and go up and down in the right places, the tune will often still be recognizable even if you get the sizes of the jumps wrong. And I don't mean slightly wrong. You could double the size of some of the jumps or make all the jumps the same size and the tune remains identifiable in many cases.[8] And thus was the drunken sing-along invented.

If, however, you keep the pitches correct and mess about with the rhythm, it's fairly easy to make even a well-known tune unrecognizable. Many tunes become difficult to identify if you simply employ the trick of making all the notes the same length. This wouldn't work too well with "Baa, Baa, Black Sheep" because the first four notes *are* all the same length, but try humming something with a bit more rhythm as a set of equal-length notes—"My Way," for example. It still sounds tuneful, but it's not easy to recognize. This is not something you can test on your own—you'll always recognize a tune you are trying to hide from yourself—but you can ask others if they can recognize tunes hummed with a different rhythm. (Warning: it's difficult to stop yourself reverting to the usual rhythm for whichever tune you try.)

After the first few hearings of a melody, we memorize the contour of ups and downs and the rhythms involved. The more complicated detail—the exact sizes of the pitch jumps—is dealt with by a different part of the brain after several hearings.[9] Young children find the details of the pitch jumps difficult, but they generally get the contour right, which is why their renditions of half-learned nursery songs sound weird but identifiable. So now we know that drunks and small children have more in common than just being unsteady on their feet and having a propensity for bursting into tears for no apparent reason.

Listening to unfamiliar types of music

To an enthusiast of any of the Western musical styles—from opera to heavy metal—most Indian classical music sounds rather baffling the first time you hear it. Similarly, heavy metal would be a shock to the system for someone who's only ever listened to Indian classical. When we hear a new kind of music, our brain needs to develop a tool kit of expectations to aid our understanding and increase our enjoyment. If you want to embrace a new genre, you'll need to acquire the appropriate tool kit. Fortunately this doesn't involve doing any conscious work. If you play just one typical piece of an unfamiliar type of music over and over in the background whenever you do the dishes, you'll find that, as the basic rules of the genre sink in to your subconscious, the "strange" music gets easier to understand and enjoy.

Every human culture engages in music of one form or another. There is an enormously wide variety, and we generally prefer the types of music we have been brought up with. But these predilections are not inborn; a baby from anywhere in the world could be whisked off to join another culture and would grow up preferring that culture's music.[10] Not surprisingly, we are at our most flexible as infants—and children brought up in two cultures can easily become bi-musical in the same way they become bilingual.

One investigation into bi-musicality, carried out in 2009, looked at people's ability to remember Indian and Western melodies that they hadn't heard before. Three groups of people were tested: North Americans, Indians, and a bicultural group with backgrounds in both cultures. The Americans and the Indians both remembered new tunes more easily if they were dealing with melodies from their own musical culture. The bi-musical folk, however, performed equally well for both types of music.[11] Having developed bi-musicality early on in life, these lucky people found it easy to fully appreciate two musical cultures.

Brain scans have demonstrated that your brain has to work harder when it's processing an unfamiliar type of music,[12] and as our brain doesn't like working particularly hard, we often dismiss unfamiliar music as unpleasant. But we have a choice: we can either deliberately seek out the new and unusual so it becomes easier to process, or we can avoid any type of music we find unusual. I think it's great that young people today are using technology to access far more types of music than we ever did (or could) when I was young. Nowadays it would be no surprise to find African drumming, bluegrass banjo, and Sinatra on the same playlist.

When tunes became unfashionable . . .

Some things develop and progress as time goes on. Passenger aircraft, for example, have definitely improved, decade by decade, since the 1930s.

Other things, such as spoons, don't progress. Spoons are useful as they are, and even advertising executives can't see any clear development route, which is why we don't have a spoon marketing board, or any spoonocentric research centers.

So, which one is music? Does it follow the airplane trajectory of development or the spoonish flat line?

Many years ago my girlfriend of the time gave me a crappy old CD with a heavily scratched case and no sleeve notes. You'll be glad to hear that the condition of the CD was no reflection on the state of our relationship. Tracy (a librarian) sometimes brought home CDs that the library planned to throw away because they were in too poor a state for lending to the public.

Sometime later I put the CD on as background music while doing some work in the kitchen. It was pleasant, modern-ish orchestral music written by a composer who was clearly fond of a good tune—just the sort of thing to listen to while you are trying to fix your toaster. About halfway through the second movement my attention drifted away from recalcitrant kitchen appliances... this music was beautiful... similar in mood to that famous Adagietto by Mahler (the one used in the film *Death in Venice*)... who was this composer? Was it anyone famous? Why hadn't I heard this before?

A quick perusal of the torn and scabby notes on the back of the case revealed that I was listening to the second movement of George Lloyd's Seventh Symphony. Who was George Lloyd, and why hadn't I ever heard of him? Or his music? These questions interested me for a few minutes, but faded to the back of my mind as I walked into town to look for a toaster shop.

So why was George Lloyd neglected by history?

The main outlet for new classical music in the UK is BBC Radio 3. The announcers of Radio 3 live in a marvelous world where only music is important. Their final broadcast will probably be something along the lines of "We interrupt this program to announce that World War Three has just been declared, and the human race will be wiped out in seven minutes, which just gives us time to listen to two of Schubert's shorter songs..."

Anyway, if you want to be a successful classical composer in the UK, it's vitally important to get your stuff played on Radio 3, but back in the 1960s this was a difficult prospect if your music was considered *too* tuneful. The intellectual trend of that time was that tunes in older music were fine, but tunefulness was getting a bit old-fashioned and new classical music should be more intellectually demanding. George Lloyd's music was one of the victims of this fashion. The committees who chose the new music for Radio 3 considered his melodious offerings out of date because they thought that music should be progressing from simple beginnings to a more complex future.

But music doesn't follow the aircraft model of development. Nor, on the other hand, is it spoon-oid and devoid of progress.

Music grows; it develops by accepting new ideas without necessarily rejecting old ones.

The way music has developed is rather like the way cooking progressed as people discovered and imported new ingredients and spices from other countries. Older elements aren't discarded; they are just incorporated into the growing number of options.

George Lloyd knew this, and used a combination of long-established ideas (tunes should be easy to follow) and newer techniques (harmonies can be complicated and occasionally vague) to produce his symphonies. But he and several other twentieth-century composers were rejected by radio stations and concert halls because they embraced both the past and the future. An intellectual movement had begun earlier in the century that had effectively banned tunes. And oddly enough, it all started in the city where most of the famous classical tunes had come from in the first place.

Although various clerics and politicians have tried to impose their will on composers throughout history, there was only one twentieth-century case of classical composers themselves saying, "Let's abandon this older stuff and create a new set of rules." The deluded folk who set off on this path in 1920s Vienna were led by a shiny-pated intellectual called Arnold Schoenberg, and they are collectively known as the "Second Viennese School" (the first Viennese School having consisted of Haydn, Mozart, and Beethoven). The composition style Schoenberg invented was called serialism—or twelve-tone—and it's bonkers, for reasons I will explain in a minute. Schoenberg was an established composer of normal-sounding classical music before he had his big idea, which he described with endearing modesty during a walk with his friend Josef Rufer: "I have made a discovery which will ensure the supremacy of German music for the next hundred years."[13]

In fact Schoenberg's ideas never achieved the wide popularity he hoped for because the musical system he designed was completely out of step with human psychology. Serialism itself is a musical backwater, but a quick look at the reasons why it failed can tell us a lot about the psychology of how we appreciate other types of music.

I have already described how "normal" Western music is put together: you pick your team of seven (or so) notes from the twelve different ones available. This is your key, and you use it to produce your tunes and harmonies—sometimes sprinkling in notes that aren't in the key for extra interest and also changing from one key to another whenever you want to. Within any one key certain notes will tend to be used more than others, and the strongest members of the team will also be used often, at points of rhythmic emphasis.

Put like this, it does all sound a bit predictable and boring—which is exactly what Schoenberg thought. Any normal, conventional rebel would have decided to abandon the rules and just compose whatever he wanted to hear. That's what fellow shiny-pated intellectual Igor Stravinsky did, and we all know what a success he turned out to be. Schoenberg had other ideas though. For him, ignoring the existing rules wasn't enough; he was going to invent a completely different set of rules.

Schoenberg looked at the key system, which he found so dull, and decided that the only way to avoid keys was not to pick a team of seven notes. Your team had to involve all twelve. Stravinsky and others, following in the footsteps of Debussy, were already using all twelve notes to produce atonal (keyless) music. But in any piece, some notes would be used more than others, and the music would settle into keys from time to time. But this was all too wishy-washy for Schoenberg. He wanted an absolute absence of keys, so he decided that the answer was to use all twelve notes to produce a series—or tone row.

As decreed by Schoenberg, the composer would begin working on a piece by first deciding what order he was going to use all twelve notes in.

Like this, for example: C, F#, A#, D, E, D#, G, F, G#, A, C#, B.

Having decided on your series, you then had to use it for all your tunes and harmonies. If a note had just been played in the "melody," it would have to wait its turn until all eleven others had been played—in the order fixed by the row.

Shortly after coming up with this basic rule, even Schoenberg saw that it was far too restrictive—so he created add-ons. You were, for example, allowed to use your row backwards, or upside down (i.e., if the original row went "down a tone, up a semitone," your upside-down version could go "up a tone, down a semitone").

I mentioned earlier that "rules" in music are simply observations put together at a later date by scholars who are working out what the musicians have been doing. Stating the rules beforehand, as Schoenberg did, is putting the cart before the horse, and much of the music produced by the serial techniques sounded pretty much as if it had been composed by a carthorse—although perhaps I'm being a bit harsh on carthorses here.

Howard Goodall, in his excellent book *The Story of Music,* puts it well:

> Schoenberg's theoretical rebellion, which later acquired the labels "serialism" or "atonality," produced decades of scholarly hot air, books, debates and seminars, and—in its purest, strictest form—not one piece of music, in a hundred years' worth of effort, that a normal person could understand or enjoy.[14]

A defender of strict serialism might say that if people were exposed to more of it, they would start to understand it and love it more. But the problem is deeper than that. Serialism presents us poor humans with a number of intractable psychological difficulties.

First of all there's the difficulty of remembering the "melody" of the row. All the world's music systems involve having only about seven notes in play at any one time. (There are systems like the Indian one, which appears to have more at first glance, but a closer look reveals that there are actually about seven, although some of them have variants that are slightly lower or higher in pitch.) The clue to how many different notes could be in play at any one time was revealed back in 1956, when the psychologist George Miller wrote a landmark paper titled "The Magical Number Seven, Plus or Minus Two: Some Limits on Our Capacity for Processing Information," in which he explained that the human short-term memory can cope with only about seven different things at any one time.[15]

The "only about seven things at a time" rule for your short-term memory applies to a wide variety of things, from remembering new names at a wedding party to recalling the items on that grocery list you left at home. It also applies to musical notes. It's why, although we have twelve different notes in an octave, we generally use only a selection (or key) of about seven at a time. If you use Schoenberg's serialism technique for composing music, every melody has twelve different notes. Although you might be able to remember the contour of such a melody, the chances of accurately remembering all those up-and-down jumps in pitch are very small.

Then there's the harmony. In standard music a lot of the ebb and flow of the emotional content depends on the buildup of tension in the harmony followed by relaxation. Relaxed harmonies rely upon chords built up of notes that have strong interrelationships. Chords of this sort are plentiful in music that is based on keys but are scarce in the serial system. The serial composer can easily provide tension but can't offer the relaxation that gives the tension context.

Finally, as I said earlier, there is the difficulty in building up expectations. As you know, we gain a lot of pleasure from the frustration or reward of our expectations of what the music will do next. If we encounter a new type of music from a different culture, then we start off finding it rather difficult to deal with. After listening to the new genre for several hours spread over several days or weeks, we start to "get it." Once we have a handle on how this new music works, we can build up expectations and a new source of musical pleasure becomes available. The problem with serialism is that the rules are completely obscure from the listener's point of view. Serialism rules are easy to write down on paper, but you can't hear them in the music. So the listener can't build up a system of expectations for the genre.

It's not surprising that serialism didn't thrive. What's surprising is that any high-quality music was produced at all. Some pieces composed by the members of the Second Viennese School have become part of the general repertoire of classical music and are considered by many (including myself) to be beautiful—but these pieces always involve a relaxation of the rules. Berg's Violin Concerto, for example, succeeds as a piece of music—but wouldn't get a pass mark in a serial music composition exam because it is only loosely based on the rules.[16]

Nowadays serialism has largely been abandoned, but it has left an unfortunate legacy. A lot of modern classical composers still think that whistle-able tunes are something to be sneered at. So it's no surprise that the most popular branch of modern orchestral music is movie music, most of which is very melodic. There is, however, one branch of classical music that is thriving even though it is not melody-driven in the traditional way. Minimalist music concentrates on textures and repeated patterns to build up a hypnotic experience that can be very emotionally moving. (If you've never heard any, you could start with Steve Reich's "Variations for Winds, Strings, and Keyboards.") The rhythmic repetitions and harmonic changes of this type of music seem to produce a continuous sequence of melodies out of thin air. This is a return to the pre-serialism pattern of musical growth because the minimalist style builds on musical techniques of the past (harmonic progressions, repetition, etc.) to produce a new sound.

Tunes, plagiarism, and copyright

Tunes are the most important feature of many types of music. In fact, the melody is so important that, apart from the lyrics to a song, it's the only aspect of a piece of music that you can copyright. A tune is the legal property of the composer—and where there's property there'll be cash, and where there is cash there'll be lawyers...

Plagiarism is, of course, just another name for stealing other people's ideas and pretending that you thought them up yourself. In music, the most frequently plagiarized ideas are melodies. But the level of wrongdoing ranges from truly criminal activity to innocently imagining that you have invented a tune when in fact you have subconsciously remembered it from a film you watched last year.

As I just mentioned, melodies can be copyrighted, and you can get into legal trouble if you use a melody that is still under copyright. Even if you do so accidentally. Even if you are the Rolling Stones.

Keith Richards and Mick Jagger have a long history of writing songs together, and Keith is well aware that Mick absorbs music wherever he goes, and sometimes thinks he has written a melody when in fact he just heard it somewhere. This is usually no problem because the original source is easily spotted. But in his autobiography *Life,* Keith tells the story of what happened when he took their recently completed *Bridges to Babylon* album home. The Richards family were sitting around listening to the album and all was going well until the track "Anybody Seen My Baby?" During the chorus Keith's daughter Angela and her friend started to sing different words. Wearing a facial expression typical of a British-millionaire-rock-star-father-of-a-possibly-sarcastic-daughter, Keith asked for an explanation of this unsupportive behavior and was shocked when Angela told him that the melody of the chorus was exactly the same as "Constant Craving" by k. d. lang. With only a week to go before the album came out, Keith took swift action, and now, if you look at the album notes, you will find k. d. lang and her collaborator Ben Mink mentioned in the writing credits.[17]

On a less glamorous level, I nearly fell afoul of the lawyers myself when I was writing my last book, *How Music Works.* I had decided to use only two tunes for all my musical examples in the book: "Baa, Baa, Black Sheep" and "Happy Birthday to You." The book was peppered with references to these two songs to explain various rhythmic techniques, the size of intervals, and lots of other stuff. Various editors and friends were reading draft chapters and I was finally getting toward the end of the project when someone said, "Isn't 'Happy Birthday' still under copyright?" This seemed very unlikely to me. I hadn't realized a song like that could ever have been under copyright. But I did some research and found that this most innocent of songs was the focal point of a story of corporate greed and lawyerly behavior worthy of a John Grisham novel.

There are an awful of a lot of rumors and anecdotes about the song "Happy Birthday to You," but thankfully there is also a comprehensive, reliable piece of academic work on the subject by Robert Brauneis, a law professor at George Washington University. In his excellent paper "Copyright and the World's Most Popular

Song," Professor Brauneis covers the whole story in great depth,[18] but here is a quick review of the highlights and lowlights.

The "Happy Birthday" song was written (with different words) sometime around 1889 by a couple of American sisters, Mildred and Patty Hill, who were both kindergarten teachers.

Now, please don't make the same mistake I did by conjuring up an image of a pair of little old ladies in lace bonnets stumbling across a tune one evening while cooking supper. At the time the song was written, Patty was twenty-two years old and her sister Mildred was thirty. And they didn't just happen upon the tune. It was carefully composed as part of an experiment in the musical education of young children.

Mildred was a composer of popular songs and also a musicologist who specialized in African American music. Patty was just beginning a career in education that would eventually culminate in a professorship at Columbia University. So what we are dealing with here are a couple of young, intelligent, highly motivated specialists involved in a musicological project.

The aim of the project was to develop a collection of songs that would be musically and emotionally suitable for small children. Mildred wrote the music and Patty wrote the lyrics. They would write a draft of a song and, in Patty's words:

> I would take it into school the next morning and test it with the little children. If the register was beyond the children we went back home at night and altered it, and I would go back the next morning and try it again and again until we secured a song that even the youngest children could learn with perfect ease.[19]

The songs that came out of this work were published in 1893 as *Song Stories for the Kindergarten*. One of the songs had the same tune as "Happy Birthday to You," but the words were completely different. The song was originally composed and published as "Good

Morning to All" and repeated that phrase, with no mention of birthdays.

Toward the end of the 1800s there were no standard birthday songs in the United States because people had only recently begun to celebrate birthdays in the way we do today. But at some point after "Good Morning to All" was published, people started to sing "Happy Birthday to You" instead of the original words. In the early years of the twentieth century the song took off as the standard birthday song, and by 1937 filmmakers knew that even a short instrumental extract of the tune would tell the audience that it was someone's birthday.

The song, with its new "Happy Birthday" words, was supposedly copyrighted by the two sisters in 1935. Through the years several people have claimed that Mildred didn't write the tune, but these accusations don't hold up. None of the "identical" or "very similar" tunes the accusers go on about is actually similar, so I think it's fair to say that Mildred is the composer. But no one knows who invented the "Happy Birthday" lyric. Maybe it was Patty, maybe it was one of the kids at the school—or someone else entirely.

Nowadays "Happy Birthday to You" is the world's most popular song. So popular, in fact, that in the time it will take me to type this sentence, it would have earned the copyright holders about $2. That's $5,000 a day—or about $2 million a year.

Which is why Warner Chappell Music were so keen to keep it under copyright. By 2013 they were claiming that they'd managed to extend the copyright until 2030! Having watched *Beverly Hills Cop* three times, I have an in-depth understanding of the American legal system and, from the point of view of my advanced training, I must say that ninety-five years of copyright seems a little excessive.

Professor Brauneis made a good case that the song wasn't really under copyright, and, using his work as a starting point, Jennifer

Nelson's tiny filmmaking company (Good Morning to You Productions) started a lawsuit challenging the copyright in June 2013. And the drama doesn't end there. On April 27, 2013, a band called Rupa & the April Fishes gave a concert in San Francisco on Rupa's birthday. The live album of the concert includes a recording of the audience singing "Happy Birthday" to Rupa. Faced with a $455 license fee for "using" the song, Rupa joined the legal challenge—as did filmmaker Robert Siegel, who was charged $3,000 for using the song in one of his films. The ensuing David and Goliath court battle came to an end on September 22, 2015, when Judge George H. King declared that Warner Chappell had never owned a valid copyright on the song. There's even talk that they might have to pay back some of the millions they've raked in over the years. So now, if you want to sing "Happy Birthday" to your grandmother in a restaurant, you can do so without fear of the legal consequences.

When it's inspiration, not plagiarism

You might be familiar with the saying "Good artists borrow; great artists steal." It's a good quote, so good in fact that it has been variously attributed to Oscar Wilde, T. S. Eliot, Picasso, Stravinsky, and several others. Of these luminaries only Eliot seems to have actually written something along similar lines: "Immature poets imitate; mature poets steal; bad poets deface what they take, and good poets make it into something better, or at least something different."[20]

If you go hunting for plagiarism in music you'll find apparent traces of it everywhere. But insisting on 100 percent originality in every piece of music would be pointless and very restricting. Without some overlap in the harmonies, rhythms, timbres, and textures used, it would be impossible to create or recognize a genre. For example, a lot of 1960s Motown tracks are similar to one another, and it's this similarity that makes the genre recognizable.

Music is propelled forward as new songs and genres grow out of previous styles. It's easy to hear the seeds of Iggy Pop's 1977 hit "Lust for Life" in the first few seconds of two songs from 1966 written by Lamont Dozier, Brian Holland, and Eddie Holland: "I'm Ready for Love" by Martha and the Vandellas and "You Can't Hurry Love" by the Supremes. But this isn't plagiarism; Iggy was creating a new sound, drawing inspiration (consciously or subconsciously) from an earlier style. This sort of thing has been going on forever of course. Brahms's First Symphony was finished in 1854, thirty years after Beethoven's Ninth, but you can hear echoes of the earlier work in the later one. It's good fun spotting links between older pieces and newer ones,* and unless someone really is trying to steal someone else's work, it's all just part of the natural progression of music.

* One of my favorite older-newer pairs is Chopin's "Berceuse" and Bill Evans's "Peace Piece."

CHAPTER 11

Untangling the Tune from the Accompaniment

A lot of the emotional impact that music has on us comes not from the melody itself but from the way it is harmonized with other notes. A change in harmony can completely change the emotional power of a tune. But as the accompanying notes and the tune are all happening at the same time, why don't we get them mixed up? How does our hearing system work out that *this* note is part of the tune and *that* note is part of the harmony?

We can always spot which part of the music is the tune, even if the melody is new to us and there are lots of other accompanying notes going on at the same time.* Our ability to do this is extremely impressive when you realize that a computer would find it impossible in all but the simplest cases. The human brain outstrips the computer in identifying melody because it's extremely good at something called *grouping*.

"But what's the big deal?" you might say. "The tune's always obvious, isn't it?"

Er—no. The tune just *seems* obvious because we are so very

* If you'd like to know how we choose the accompanying chords and harmonies for tunes, you'll find more information in "Fiddly Details," section C, "Harmonizing a Tune."

good at spotting it and separating it out from the background notes even if all the notes involved have the same timbre and loudness. In a harmonized piece of music there are so many possible tunes going simultaneously that our to ability spot the one the composer had in mind is astonishing.

In fact, as we shall soon see, the number of options we have to choose from is staggering, even if you pick just a few seconds of a simple tune with a very uncomplicated harmony.

Yes—it's "Baa, Baa, Black Sheep" time again...

Let's say I have hired some flute players to entertain us by playing my latest masterpiece, an arrangement of the first line of "Baa, Baa, Black Sheep" for four flutes.

Our most expensive flautist will play the tune:

C_5 C_5 G_5 G_5 A_5 B_5 C_6 A_5 G_5.

The other three flautists will play the accompanying harmony—each of them playing one note for each of the melody notes—and they will have a fairly boring time of it because their parts involve lots of repeating notes:

Flautist 2 will play: G_6 G_6 G_6 G_6 F_6 F_6 F_6 F_6 G_6
Flautist 3 will play: E_4 E_4 E_4 E_4 C_5 C_5 C_5 C_5 E_5
Flautist 4 will play: C_4 C_4 C_4 C_4 A_4 A_4 A_4 A_4 C_4

Now, you might think that the number of tune options our brain would have in this case would be very small. After all, the piece is only nine notes long.

But there are more melodies hidden in this simple piece of music than you could possibly imagine. This is because any sequence of notes can be a tune. In fact flautists 2, 3, and 4 are all playing their own tunes. Dull, repetitive tunes it's true—but they are tunes.

So, how many hidden tunes are there?

The flautists as a group have a job to do. They have to deliver a sequence of nine groups of four notes to the listeners:

Group	1	2	3	4	5	6	7	8	9
	C_5	C_5	G_5	G_5	A_5	B_5	C_6	A_5	G_5
	G_6	G_6	G_6	G_6	F_6	F_6	F_6	F_6	G_6
	E_4	E_4	E_4	E_4	C_5	C_5	C_5	C_5	E_5
	C_4	C_4	C_4	C_4	A_4	A_4	A_4	A_4	C_4

In my original arrangement I happen to have given a certain selection of notes to flautist 1 and another selection to flautist 2. But if they swapped a few notes, the audience wouldn't know the difference. In fact we could let all four flautists divvy up the notes from each group in whatever way they liked (before they started playing), and as long as all the notes were played in the correct sequence, the sound of the music wouldn't change as far as the audience is concerned.

But when the flautists divide up the notes among them, they are choosing new tunes for themselves. And because our flautists have nine groups of notes, each with four choices, the actual number of possible tunes in our simple example is four multiplied by itself nine times: 4 x 4 x 4 x 4 x 4 x 4 x 4 x 4 x 4 = 262,144. Over a quarter of a million tunes* are hidden in just four seconds of music! If they tried all of these possibilities, our flautists would be swapping notes for several weeks before they got to the end of the project.

If we did this same exercise for the thirty-eight notes of the whole tune (with three notes of accompaniment for each note in the melody), the number of possible tunes would be over seventy-five thousand billion billion, which is more than the number of grains of sand on all the beaches on earth.

---

* If you'd like a bit more explanation about how we get to this huge number, please read "Fiddly Details," part D, "How Many Tunes Are Hidden in the Harmony?"

For a pop song, or any piece of harmonized music a few minutes long, the number of possible tunes would be more than the number of stars in the universe!

And these possible tunes are not something I've made up; they really are there. It's just that we ignore the billions of possibilities and hear only one. Given the vast number of options, I think you'll agree that our effortless ability to spot the "real" tune—the one the composer had in mind—is pretty amazing.

We achieve this phenomenal feat by using a collection of pattern recognition techniques that are part of our mental survival equipment. Our brain is continuously bombarded with vast amounts of data from our senses, and we simply wouldn't live very long if we tried to analyze every detail. For this reason we have found it useful to group the things we see and hear into categories.[1] Your average caveman didn't stand around wondering if the unfamiliar spotted snake slithering toward him was as dangerous as the stripy one that killed his brother; he'd automatically put it in the "Snake! Run like hell!" category. Our categorization system groups things in several ways, and here are four of the main grouping methods, together with examples of how they are useful for survival:

Similarity. If things have features in common, we categorize them together. If we see an unfamiliar twenty-foot-tall plant with leaves and branches, we don't need to worry about exactly what it is. We can simply categorize it as a tree. Having done so, we can assume that it will have tree-like qualities, which means that we can probably burn pieces of it to keep our fire going.

Proximity. If things are near one another, we categorize them together. You are using this skill at the moment to group letters into words as you read this sentence. If we find that a particular red currant bush down in the valley provides juicier berries than the bushes on the hillside, it's likely that the other red currant bushes in the valley will also have high-quality fruit.

Good continuation. If several images look as if they are part of an incomplete group that makes up a bigger picture, your brain tends to join up the dots. This skill allows you to work out what you are looking at even if your view is obscured by something. This is what helps us to see a snake hiding in the branches of a tree.

Common fate. We have an ability to identify if several things are traveling in the same direction toward a certain end point. Watching a flock of birds in flight, we immediately notice if a few of them start to veer off in a slightly different direction. If this happens, we categorize them into two separate groups with different common fates. Common fate lets us identify which antelope is getting left behind by the herd, and in which direction the bees are buzzing back to their honey-filled hive.

These innate skills—like all innate skills—are part of our equipment for staying alive, and they allow us to sit down in a snake-free restaurant and pour red currant jus over our smoked venison fricassee.

OK—so what's all this got to do with music?

Well, remarkably, our hearing system untangles the tune from the harmony by using the four skills I've just discussed.

Similarity. In many cases the instrument carrying the tune has a distinct sound or timbre. Often it's a human voice, but in other cases it might be, for example, a saxophone or guitar. We link sounds with a similar timbre together, and this helps us follow the path of a tune. The accompanying instruments (e.g., keyboards or violins) have their distinctive timbres as well, which also helps us keep the harmony separate from the melody.

Proximity. There are two ways in which proximity helps us follow the melody. One is proximity in time: the notes of a melody follow one another in a sequence, with one note starting as the previous one ends (with occasional gaps). Because the notes are generally nose to tail like this, we can easily identify the individuals as being parts of a chain (the tune). Proximity in pitch is also an important clue that helps us identify the tune, because most of the time tunes use only small steps in pitch between consecutive notes.

Good continuation. A tune is always on its way somewhere—up toward its highest note or down toward its lowest, with diversions along the way. In many cases we can guess what note will be next, and even when we can't forecast the exact note, we can be pretty confident that certain notes are extremely unlikely. The way we keep track of the tune is impressive because in almost any music there will be accompanying notes that are timed to occur between the melody notes. To make a visual analogy here, the accompaniment will be interfering with our view of the melody. And the melody will be obscuring our view of the accompaniment. (The snake hides some of the branches and the branches hide some of the snake.) Our good continuation skills keep the melody separate by showing us that any particular accompanying note is very unlikely to be part of the melody and similarly that the next melody note probably isn't part of the accompaniment. We use "good continuation" assumptions when listening to music in the same way that we separate out superimposed images on paper. For example, good continuation makes us interpret this image as a triangle within a circle—not as three capital D's touching corner to corner.

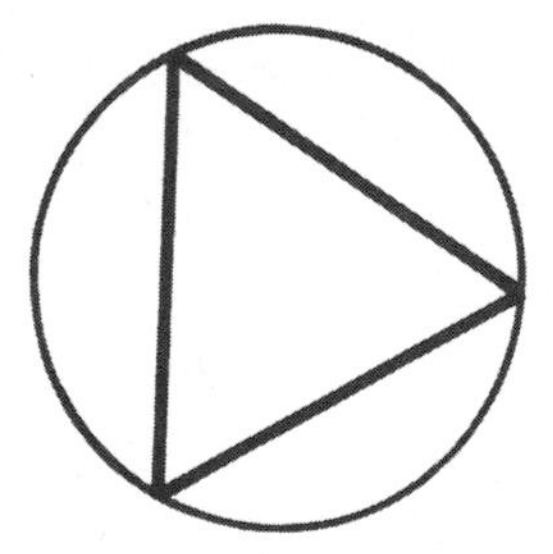

Common fate. If the music we are listening to is being played by a group of similar instruments (a brass band or string orchestra, say) or by a solo instrument (a piano or a guitar), then the timbre of the melody will be the same as the timbre of the accompaniment. In this case one way to keep the tune separately identifiable is to pitch it much higher than the harmony. But eventually this will get boring and in many cases the tune will eventually descend back into the same range of notes as the harmony. You might think that we'd lose track of the tune in this situation, but fortunately this is where our ability to identify common fate comes in handy.

I mentioned earlier that the melody is always on its way somewhere—but the harmony also has its own agenda. The harmony is generally less dynamic and more obviously organized than the tune, and again, we can use a visual analogy here to show how we keep the tune and harmony separate. There is a cliché in action films whereby a couple of people who are chasing each other suddenly come across a parade of people marching down the street. At this point the person being chased disappears into the marching crowd and the chaser usually shinnies up a lamppost to look for him. Quite often the fugitive is spotted easily because he's not traveling in exactly the same direction as the main crowd. The steady movement of the crowd makes it easy to identify the contrary motion of the individual. In this analogy the marching crowd is the harmony, the melody is the fugitive dodging through the crowd, and you are stuck halfway up a lamppost watching what's going on.

Songwriters and composers automatically make use of the listener's categorization skills because they are using their own when they write the music. In most cases the music consists of a tune and an accompaniment, and by now I hope you're impressed with your brain's ability to untangle one from the other. But there's more... In many types of music there are occasions when more than one tune is playing at any one time. This technique, which is called *polyphony,* was particularly popular around 1700, and J. S. Bach was the master of the craft of having two or more tunes going at once, for several minutes at a time. Polyphony is also used, generally only for a few seconds at a time, in folk, pop, rock, and jazz music. For example, if you listen to the Beatles' "Hello, Goodbye," there is a section halfway through the song where two sets of lyrics are being sung, each with its own melody, at the same time. You can also hear Simon and Garfunkel using the same technique in their version of "Scarborough Fair." In most modern music polyphony is used only sparingly, to add depth and interest to the sound. Under the main melody a short tune might be given to, say, the bass guitar, which will then pop out of the background for a little while before rejoining the accompaniment. An example of this can be found in the Rolling Stones' version of "Route 66." About forty-five seconds into the song, as Mick Jagger sings the few words that end with "pretty," the bass plays a rising "melody" of four notes before returning to more standard accompaniment.

Our abilities to identify similarity, proximity, good continuation, and common fate work simultaneously to help us make sense of what we are hearing, even if there are several tunes going on at once. We also rely on our natural tendency to look for patterns we are already familiar with to help us follow the tune. Any Western music that you hear for the first time is full of little segments you've heard before in other pieces. For example, most tunes will contain fragments of scales, and lots of familiar intervals. Our mental library of these building blocks helps us make sense of new pieces fairly rapidly.

So listening to music requires a lot more mental juggling than you might have expected. Thankfully, all the information processing is subconscious and effectively effortless, so it doesn't interfere with our enjoyment.

CHAPTER 12

Don't Believe Everything You Hear

You'll probably be astonished to learn that most of the music we listen to is out of tune, even when the musicians involved are experts. There are three main factors involved in this pitch sloppiness:

1. Our main scale system deliberately makes use of notes that are all slightly out of tune.
2. Musical instruments drift out of tune quite easily.
3. Musicians make small errors all the time.

Thankfully our senses are designed to deal with inaccuracy, and in general, we blithely overlook any errors and get on with enjoying the music. In this chapter we'll be taking a look at how our brain deals with all the out-of-tune-ness. But before we start, we need to understand what it means to be "in tune."

What does "in tune" mean?

"In tune" generally means that the notes involved sound good and smooth when they are played together. For a perfect example of in-tune-ness, let's take two notes an octave apart. As we saw in chapter nine, if you play a note simultaneously with the note an octave above it, the two notes mix together excellently:

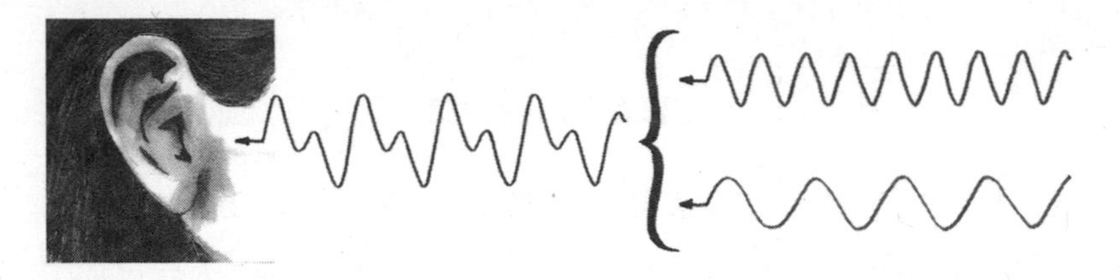

This happens because the cycle time* of the higher-pitched note fits precisely twice into the cycle time of the lower note.

But an octave is a very large interval. Most of us have a singing range of only about two octaves. So to get a workable music system with several notes in it, we need to divide the octave into smaller steps. We need to create a scale that starts on one note and rises up to the note an octave above it. And as far as possible, we want all the notes in our scale to be in tune with one another.

We'll begin by calling the lower note the keynote.

And now we have a bit of a conundrum on our hands. How are we going to choose the notes in our scale?

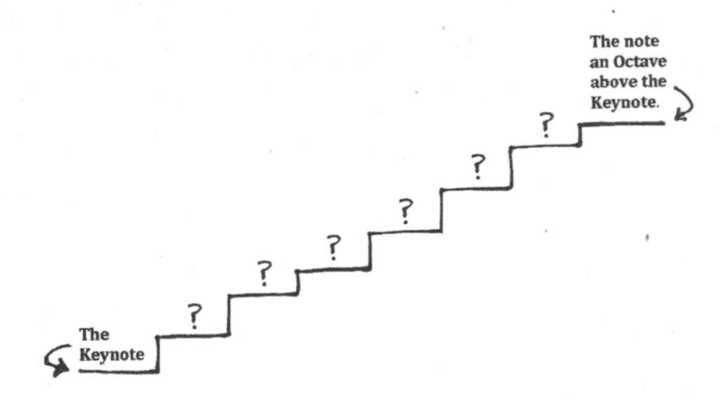

* As I explained in chapter nine, the cycle time of a note is the amount of time it takes to complete one cycle of its particular air pressure ripple pattern. The number of these cycles that can fit into a second is the frequency of the note. If a note has a cycle time of one hundredth of a second, then its frequency is 100 Hz.

Initially we have only two notes in our scale: the keynote and the note an octave above it.

And as I just mentioned, these two notes are in tune with each other because two halves are equal to one. So perhaps we should look at other simple fractions. Maybe a note that has two thirds of the keynote cycle time might be somehow in tune with the keynote?

And yes, this note and the keynote do sound good together. Two cycles of the keynote match three cycles of the new note to give a nice repeating pattern.*

Because the repeat is spread over two cycles of the keynote, it's not quite as strong a link as the one between two notes an octave apart (you can still hear the two separate notes), but we nevertheless hear it as a very pleasant combination. You can test this for yourself if you've got a friend nearby.

In "Baa, Baa, Black Sheep," "baa" is the keynote and "black" is the "⅔ keynote cycle time" note. So, sing "Baa, baa" and make the second "baa" last several seconds. In the middle of your long "baa," get your friend to sing "black" at the correct pitch and you'll hear how well the two notes go together. (If your dog starts yowling you're probably doing it wrong.)

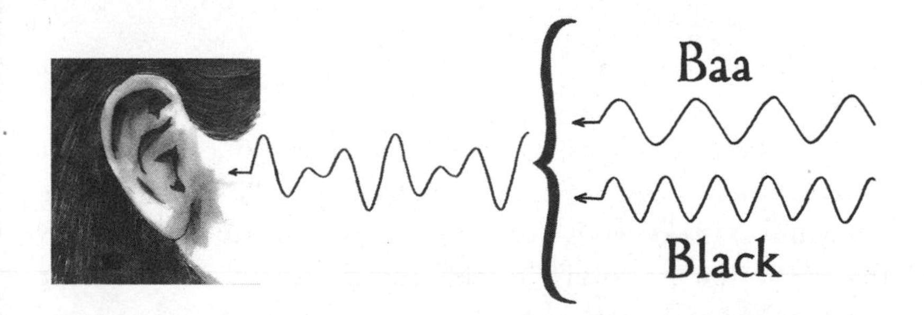

* You'll find a bit more detail about how scales are put together in "Fiddly Details," section E, "Scales and Keys."

Carrying on with our simple fraction logic, if we look at the note that has three quarters of the keynote cycle time, we can see that four of its cycles fit exactly into three of the keynote cycles. And sure enough, this does make another pleasant-sounding combination. If you and your friend are feeling in particularly fine voice today, you can try this one using the first two notes of "Here Comes the Bride." (You sing a long "here," your friend joins in with "comes.")

We can create a major scale of eight notes that are all as in tune with the keynote as they can be by using simple fractions like this. Here are the fractions of the keynote cycle time for them all, starting with the keynote:

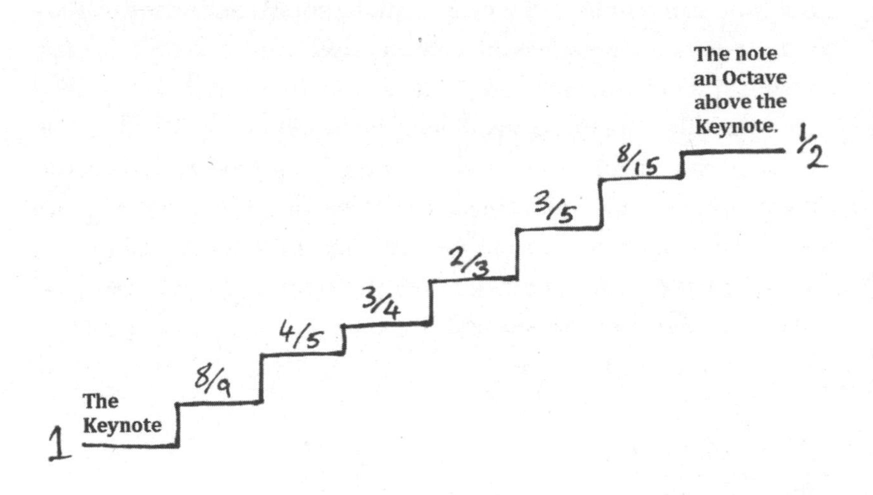

This system of choosing notes from simple fractions creates what is called a "Just" scale. If you play any of these notes at the same time as the keynote, they set up a repeating pattern similar to the one I drew for "Baa"/"Black," and that's how two different notes can be in tune with each other. The smoothest-sounding combinations come from the fractions with the lowest numbers above the line. So the keynote and the ½ note (the octave) sound smoothest together. The keynote with the $\frac{2}{3}$ note is the next-smoothest combination, then the keynote with the $\frac{3}{4}$ note, and so on. The final two fractions ($\frac{8}{9}$ and $\frac{8}{15}$) sound rough when we play them with the keynote because they involve too many (8) keynote cycles before the pattern repeats, but we need notes in these positions; otherwise we'd have big gaps at either end of our scale.

So, our definition of "in tune" goes like this:

If a few cycle times of one note fit exactly into a few cycle times of another note, the two notes are in tune with each other.

And it would be lovely if the Just scale obeyed this rule all the time. But unfortunately, even though the notes of the Just scale are all in tune with the keynote, they are *not* all in tune with one another. For example, if the $\frac{8}{9}$ note and the $\frac{3}{5}$ note are played together, they sound unpleasantly out of tune because the two fractions are incompatible with each other. If a couple of singers or violin players had to play these two notes together, one of them would have to change the pitch of their note a little until the combination was in tune. But it's impossible to do this sort of thing on an instrument like a piano, where the pitches of the notes can't be changed while you are playing.

Basically, you can't use the Just scale system on any fixed-note instrument* because some combinations of notes would be out of

* Fixed-note instruments (keyboards, harps, marimbas, etc.) have all their notes tuned to the equal temperament (ET) scale, and the pitch of the notes can't be changed by the performer. On a completely variable-pitch instrument (violin, cello, slide trombone, etc.), you can play any pitch you like within the range of the instrument. On partially variable-pitch instruments (flute, saxophone, guitar, etc.), the player has a limited ability to alter (bend) the pitch of the standard notes for the instrument while playing.

tune, and some entire keys would be more out of tune than others.*

To overcome this problem, a different scale system called equal temperament (ET) was developed. If you'd like more detail about this and the Just system, please have a look at "Fiddly Details," section E, "Scales and Keys." But for now all we need to know is that the equal temperament system corrects the occasional serious tuning problems of the Just system by making all the fractions slightly wrong. So now all the notes (except the octaves) are slightly out of tune with one another. This means that all the piano music you've ever heard has been slightly out of tune.

In fact, most of the pieces of music you've ever heard on any instrument will have been slightly out of tune for one reason or another. But even if we had a tuning system that worked perfectly, you couldn't trust musical instruments to stay in tune.

Inaccurate musical instruments

Let's imagine we are off to a concert of flute music in the local concert hall. The flute in question is an expensive piece of engineering that has been tuned to the equal temperament scale as accurately as possible. So what could possibly be wrong with the notes it produces?

Well, for a start, let's hope that the flute player is not a beginner, because nearly all the notes on a flute are out of tune. When I said "tuned as accurately as possible" in the previous paragraph, I omitted to tell you that, even if we ignore the flaws in the ET system,

---

* For example, if you organized your Just tuning around the note of C, then most of the key of C major would work out very nicely, but the key of F sharp major would have lots of mismatched combinations that would sound out of tune.

the physics of the instrument prevents us from putting the finger holes in the correct places if we want both the high notes and the low notes to be perfectly tuned. So the hole positions are a compromise. To compensate for this, expert flute players spend a lot of time learning to change the position of their upper lip during any performance in order to alter the pitch of various notes.

Other wind instruments have similar problems. On clarinets and saxophones the compensations need to be made by increasing or decreasing the pressure of the lips on the reed. And here is what John Backus, a well-known author on the subject of musical acoustics, has to say about the bassoon:

> The bassoon demonstrates woodwind deficiencies especially well. Practically every note of this instrument is out of tune and needs to be pulled into tune with the lips.... Except for having keys added, it has changed little in several hundred years, has aptly been called a "fossil" and badly needs an acoustical working-over. (The author is a bassoonist and hopes to do this working over if he lives long enough.)[1]

Unfortunately, Dr. Backus died in 1988 without completing the project, and the bassoon is still an acoustic box of frogs.

But at least players of wind instruments can console themselves with the fact that their prehensile lip manipulation skills make them the best kissers in the orchestra.

To return to our flute concert, the hall is warming up as we come to the final few pieces. When we came into the hall, the temperature was about 60° F (15° C). The body heat of the audience has heated up the room until it's now 80° F (25° C). In this warmer air the sound waves bouncing up and down inside the flute travel faster and all the notes are now about one third of a semitone higher than they were at the beginning of the concert.

And tuning inaccuracies aren't, of course, restricted to wind instruments. How often have you seen a guitarist retune a string or two between songs? And how often did you notice that the guitar was out of tune during that last song?

The note given off by a guitar string is related to the tension on the string and its length. Guitar strings are made of steel or nylon, and the body of the guitar is usually made of wood. Steel, nylon, and wood all expand by different amounts if you warm them up by taking them from a cold backstage area onto a hot concert hall stage, so the tension on the strings will change and the guitar will go out of tune. This is why, when you go to a concert, all the guitars are already onstage, and they are often being tuned by the roadies until just before the band comes on. Later, as the gig gets really hot and sweaty, the guitars will need retuning. Also, the constant twanging of the guitarists tends to stretch the strings—so they get slacker and drop in pitch. It's a busy life being a guitar string.

Musicians are only human

As we've seen, it's difficult to keep most instruments in tune. In fact the only instrument that stays in tune all the time is the electronic keyboard. Keyboards also don't allow musicians to make mistakes in the pitch of the notes they play. On a full-sized keyboard you simply get eighty-eight different notes all tuned to equal temperament (so they are all slightly out of tune with one another) and that's it. By contrast, musicians who sing, or who play completely variable-pitch instruments (violins, cellos, slide trombones etc.), have lots of freedom to choose the pitches of their notes, but this means that it's very easy to make mistakes and produce notes that are higher or lower in pitch than the ones you intended.

The cello demonstrates this problem most clearly because the notes can be a long distance from one another. Sometimes a cellist will have to move her hand a foot or so at karate-like speed and try to stop as accurately as possible. In most types of music the notes need to be played smoothly, one after the other, with the minimum amount of silence between them. Composers would really like a zero gap between two notes, but that would give you zero time in which to move your hand. In reality a professional player will often have to complete fast, large, accurate movements in a tenth of a second or less!

Cellists do this sort of thing all the time, and it involves an awful lot of skill because there's nothing on the neck of the cello to tell you where the correct positions are—you just have to learn. Which is why it takes a long time to train to be a good cellist. As a beginner you wouldn't be doing big jumps, but if you tried it, you'd often overshoot or undershoot by half an inch or more—which could make you about a semitone out—and you'd end up producing the wrong note altogether.

Obviously, as you train you get more accurate, but even after twenty years of practice, the chances of a human executing a lightning-fast hand movement of a foot or so with an accuracy of, say, a twentieth of an inch are vanishingly small. So just how accurate is a professional cellist?

Neuroscientist Jessie Chen and her team developed a specially built cello that measured exactly where a cellist's fingers landed on the neck of the instrument while it was being played. They tested eight highly trained cellists and found that when they were moving rapidly between two notes nearly a foot apart, even professionals were often more than a quarter of an inch off.[2]

As part of the investigation the cellists were also asked to play a tune that, at some point, involved an ascending phrase of three notes, each one a semitone higher than the previous one. The middle note of the three had a very short duration: it was supposed to last only one eighth of a second (which is about as long as it takes to pronounce the "a" in the phrase "of a second"). To both the scientists and the cellists the tunes sounded as if they were played accurately, but analysis of the cellists' finger movements told quite a different story.

Actually, what typically happened was that, during the eighth of a second, the cellist's finger started at about the right place and then slid about a third of an inch during the production of the short note. The resulting blur of sound didn't actually have a fixed frequency, but it sounded fine. Fortunately, in any fast, simple passage of music like this, the listener—and the instrumentalist—pay most attention to the notes at the beginning and end of the phrase. So we don't generally notice whether or not the intermediate notes are blurred or inaccurate.

But don't sack all the cellists. Given the times and distances involved, they should be praised for getting the notes even approximately right.

So, the music we hear is often out of tune. The scales we use have intrinsic tuning problems, most of the instruments aren't perfectly accurate, and musicians make mistakes.

With all these problems, how does music still sound OK?

How our brain makes sense of all this pitch sloppiness

We manage to enjoy music even though a lot of it is slightly out of tune because our senses are champions of approximation and the editing of information. Your senses don't need to be completely accurate; it's more important that they are very fast, and accurate *enough*. You are continually flooded with all sorts of information, and your brain has to process the important bits rapidly to come up with a meaningful conclusion, such as "Oh, look—a carrot."

And if you think that's the last we've heard of carrots in this tightly woven narrative, you are woefully mistaken.

Because your senses are all you have, and they do such a brilliant job, it is tempting to think that they are flawless. Let's take our eyesight as an example. We all know how easy it is to confuse the eye with dodgy information like this image:

Even if your eyes aren't being fed deliberately misleading information, they can make mistakes—that is, they can "see" things that aren't there or completely miss things that are right in front of you. On the other hand, your vision system can also be amazingly good at filling in gaps in the information it receives. For example, each of your eyes has a blind spot that is surprisingly close to the center of your field of vision. Light enters your eye through the iris, and the image of what's going on is projected onto the back of your eye like a movie being projected onto a screen. But near the middle of our "screen" there is a hole where the optic nerve connects with the retina to take all the information off to the brain. You never notice this blind spot unless you look for it. But try this:

Close your left eye and extend your right arm as far as you can, with the thumb pointing upwards. Your thumb should now be at arm's length, level with and in front of your nose. Swing your arm slowly to your right while staring directly forwards. Don't follow your thumb with your eye. Keep your right eye stationary by focusing on something directly in front of you, and keep your left eye closed. When your thumb is just a little bit to the right of your right shoulder it will disappear!

Yet on a day-to-day basis we never realize that there is a blind spot.

Your vision system, like your hearing system, works in three stages. First, information is collected, then it's analyzed, and finally a summary is presented to your conscious mind. When we look at the impossible triangle we saw earlier, the data is collected and analyzed correctly, but the third step in the process, the summary, breaks down. In the case of the blind spot, we have an error at the first stage—the data collection. Fortunately the second and third stages are so sophisticated that they usually compensate for any gaps in the information and construct a useful summary anyway: "Yes—that's a bunch of carrots all right."

Just like your eyesight, your hearing system can be fooled, but usually it does a great job of producing useful summaries from inaccurate or incomplete information. The rather sloppy give-and-take of our hearing system allows us to understand and/or enjoy what's going on far more rapidly that we could if we always required perfect, accurate information.

Categorization of notes

Probably the most important function of our hearing system nowadays is that it allows us to understand what other people are saying to us. There are hundreds of accents, affectations, and afflictions that affect how we pronounce words—yet we manage to understand one another surprisingly well. If you encounter someone with a strong unfamiliar accent you might have a bit of difficulty, but it's rarely enough to prevent you from communicating.

Computer-aided sound analysis has shown that spoken sounds can be broken down into smaller components, each of which plays a role in determining what we hear. Detailed investigation in the case of *d* and *t* has revealed that the main components of the two sounds are very similar, but we use them slightly differently depending on which letter we are pronouncing. For a *d,* the two main components of the sound begin at almost the same moment, but for a *t,* one component starts about six one hundredths of a second later than the other.[3]

OK—so we can get our computer to make *d* noises (as in "drip") by producing the two component sounds together, then we can start to delay one of the components by one hundredth of a second, then two hundredths, then three and so on, up to six. You might think that we would start off hearing a clear *d* that would gradually become less distinct until it eventually turned into a *t* (as in "trip") once the delay was close to six one hundredths of a second—so we would hear *d, d?, ?, ?, t?, t.*

But that isn't what happens. Tests have shown that we don't hear a sliding scale between the two sounds.[4] We hear a *d* until a certain delay is exceeded and then we flip over to hearing a *t.* We hear *d, d, d, t, t, t.*

This is because we don't find a vague noise somewhere between a *d* and a *t* useful, so we use our categorization skills to divide all the *d/t* sounds into two categories—either *t*'s or *d*'s. This is inaccurate but efficient. Communication would be almost impossible if our brain accepted only an exact sound as a *t* or a *d* (or any other verbalization). We need to quickly put sounds that are approximately OK into the correct category so we can work out what is being said. Our senses do this useful approximation thing a lot. We do it, for example, when we're identifying whether something is a carrot or not. We accept that the particular orange pointy thing we are looking at is a carrot even if it's bigger or smaller than average—because it still fits into our mental carroty category.

"And what," you might reasonably ask, "has all this to do with music?"

Well, we use categorization techniques all the time when we listen to music. If a note is slightly out of tune, we don't usually notice it. We don't require the note to be exactly the correct frequency; we just need it to be close enough to be put into a category.

Back in 1973 Simeon Locke and Lucia Kellar decided to find out how out of tune a note could be and still be allowed into its category. They played three-note chords to musicians. The outer two notes of the chords were in tune, but the middle note varied from in tune to out of tune. The experimenters used the range from C to C sharp (the next note up) for the middle note. Sometimes it was an in-tune C, sometimes it was an in-tune C sharp, but most of the time it was somewhere between these two notes—and therefore out of tune.

My forefinger is on C sharp, my second finger is on C. If you (like Locke and Kellar) want to produce a note between these two, you need to get a piano tuner to come in and slacken the C sharp strings a bit.

The musicians had a properly tuned chord played to them first and were then asked if the next chord they heard (which was usually out of tune) was the same. If their hearing systems were entirely accurate, they should have spotted that the middle note of the chord was now out of tune. But generally they categorized whatever they heard as being *either* a C or a C sharp.[5] So in music as in speech, our hearing system is rather forgiving and inaccurate.

We are, however, more sensitive to notes being out of tune with one another in certain circumstances than we are in others. It all depends which notes in the chord are involved.

I mentioned earlier that, when played at the same time as the keynote, the $\frac{1}{2}$ cycle time note sounds smoother than $\frac{2}{3}$, which sounds smoother than $\frac{3}{4}$, and so on.

One odd feature of this hierarchy of smoothness is that the smoothest combinations are the ones that are most negatively affected if one of the notes is out of tune with the other. When played at the same time as a keynote, an out-of-tune $\frac{1}{2}$ cycle time note (octave) sounds the worst, the next worst is an out-of-tune $\frac{2}{3}$, then an out-of-tune $\frac{3}{4}$, etc. By the time we get to the $\frac{8}{9}$ or $\frac{8}{15}$ notes, we barely notice if they are out of tune when they are played at the same time as the keynote because the combination is so rough-sounding even when they are in tune.

Think of it like this: we don't notice an extra smudge on an already dirty mirror, but a fingerprint on a clean, polished mirror stands out a mile. Simeon Locke and Lucia Kellar knew that if they de-tuned the $\frac{2}{3}$ note in their chord, everyone would notice it pretty quickly. So they de-tuned the other note in the chord—the one with a cycle time of $\frac{4}{5}$. They were then starting with a "dirty mirror" and making it dirtier by de-tuning it. During the experiment they varied the $\frac{4}{5}$ note from a ratio of $\frac{4}{5}$ to $\frac{5}{6}$. When the note was $\frac{4}{5}$, the musicians heard a major chord. When it was $\frac{5}{6}$, they heard a minor chord. During the experiment the note was usually between these two fractions and therefore out of tune. But the categorization systems of the listeners ignored the out-of-tune-ness and only heard either a 4/5 or 5/6 note.

Another experiment to investigate our sensitivity to out-of-tune-ness was carried out in 1977 by psychologists Jane and William Siegel, who were trying to find out how accurate we are in identifying intervals. (An interval is the size of the jump in pitch between two notes.)

The Siegels pre-screened twenty-four highly trained music students, all of whom played several instruments and had begun their musical training at an early age, to determine how accurate they were at identifying intervals. Only the best six were chosen to go on to do the test.

These six students were then asked to identify a series of intervals and, at the same time, judge how in or out of tune they were.

Only three of the thirteen intervals used in the tests were actually in tune, but during the test all the students insisted that eight or nine of them were correct. The students thought that the intervals were correct even if they were out by one fifth of a semitone. A computer could have shown them how far off the notes were—but everyone involved subconsciously fitted the intervals into categories using our relaxed mental approximation techniques.

The results of the tests were a triumph for the psychologists, who had been expecting categorization to win the day. But it was all a bit embarrassing for the music students, who assumed that they would be much more accurate than they actually were.

As the Siegels put it in their report on the experiment:

> Most were unaware of how poorly they were doing during the course of the experiment, and considerable tact and compassion were required when we debriefed the subjects.

And the final verdict:

> Musicians had a strong tendency to rate out-of-tune stimuli as in tune. Their attempts to make fine judgments were highly inaccurate and unreliable.[6]

Given all this mental sloppiness, it's rather surprising that our ears can actually distinguish between notes that are *very* close together in pitch. Well-trained listeners can tell that two pure tones* are different even if their frequencies are as close together as 1000 Hz and 1002 Hz—a difference of only about one thirtieth of a semitone.[7] When the researchers tried untrained listeners with this pitch discrimination test, they were far less sensitive at first, but they caught up with the experts after only a few hours' practice.[8]

But for both our vision and hearing systems, there is a big difference between being able to distinguish between two things and being able to remember the difference between them.

No doubt you've all seen those paint color charts we pore over before we decorate the bedroom—although, frankly, I don't know why the paint manufacturers bother with all the bright colors nowadays, since nearly everyone I know chooses variations on a hue that should be called "damp string." There is usually a selection of about a dozen of these beiges, and countless newlyweds have wasted their entire honeymoon quibbling about the merits of decorating their new home with "Bamboo Dream" as compared to the even subtler tonal underplay of "Cardboard Dawn."

Well, imagine if there were a choice of fifty beiges rather than a dozen or so. You would have no trouble distinguishing between them on the color chart—but could you remember your favorite shade without looking at the chart? Let's imagine a bizarrely motivated burglar broke into your house and cut up your beige color

---

* Pure tones are computer-generated notes with a very clean (though rather boring) sound. For more details, see "Fiddly Details," section A, "Timbre."

chart into fifty little nameless squares and then stole five of them. You could stare all day at the forty-five remaining squares, but you'd never be sure that your favorite wasn't one of the missing ones.

Or, to consider another implausible scenario: I could show you three pieces of string—a long one, a short one, and a medium-length one. If I presented you with one of them in isolation later in the evening, you could easily remember whether it was the long, medium, or short one. But let's say I gave you twenty pieces of string. You would still have no problem in ranking them in order of length. In this case, however, if I presented you with one of them in isolation a couple of hours later, you wouldn't be able to tell me exactly which one it was. You might be able to remember that the one I was showing you was somewhere in the middle of the range, but you wouldn't be able to identify whether it was the eighth-, ninth-, or tenth-longest. We are all far more skilled at distinguishing between things than we are at remembering the details of the items involved. This rule also works for music. Our ability to remember tunes depends upon the fact that there are only a few notes involved, and they have fairly large jumps in pitch between them.

With only seven fairly widely spaced notes to choose from in any key, we can use our categorization skills to make sense of and remember tunes even if the musician or instrument is a little inaccurate in producing the correct note pitches.

Singers, and players of completely variable-pitch instruments (violins, cellos, slide trombones etc.), can decide on the scale system they will use depending on the context. If they are playing along with a piano, they might choose to match its ET tuning. If they are playing solo or with a bunch of other variable-pitch instruments (in a choir or string quartet), they might choose to play according to Just tuning.

Whichever they choose, they are often somewhere between the two (or off target altogether) because of the natural difficulties in hitting exactly the correct pitch—particularly when playing quickly. For example, on a violin neck the difference between the ET note and the Just one is often only a tenth of an inch or less. Even professional players are not that accurate when they are playing even moderately quickly. This works out quite well because our hearing isn't accurate for short, rapid notes either. So neither the instrumentalist nor the listener is aware of the errors in pitch.

The only time when small errors in pitch are really troublesome is in harmonies at moderate to slow speeds. When we hear long notes played together, we can spot if they are out of tune, particularly if the out-of-tune notes are the prominent ones in the chord. In this type of music, whether it's a Beethoven string quartet or a harmonized gospel song, errors in tuning stand out very clearly. Fortunately these are also the occasions when errors can be rapidly repaired. Skilled performers on variable-pitch instruments (including the voice) will quickly and instinctively adjust the note they're producing to bring it into tune with the others that are playing at that point. In this way all the peculiarities of the Just system can be ironed out as you go along.

Some players of variable-pitch instruments, playing the melody against a background of harmony, or playing completely solo, use *expressive intonation*. This involves making some of the jumps in pitch in the tune slightly bigger and some of them slightly smaller in order to emphasize certain notes in a tune. For example, some melodies end with the keynote preceded by the note just below it. Playing this penultimate note at a slightly higher pitch than normal exaggerates the feeling that we are on our way home to the keynote, and we're nearly there. Blues guitarists and singers use similar techniques, sliding up toward notes and then away from them (in some cases never reaching the "proper" pitch of the note). Supporters of expressive intonation and blues techniques say that they add more oomph to the emotional content of the music. Detractors say they merely sound out of tune.

* * *

Given all the approximations and inaccuracies involved in every type of music, we should be thankful that our ability to categorize helps us make sense of it all. It's amazing how well everything works out. But no matter what tuning system we use, or how accurately we use it, there will always be notes that sound rough when they are played together, an effect known as dissonance.

CHAPTER 13
Dissonance

Dissonance is the word we use to describe unpleasant combinations of notes. There are two sorts of dissonance in music. One type, *sensory dissonance*, is based on the science of how your hearing system works. *Musical dissonance,* by contrast, is more concerned with your musical taste. These are completely different concepts, so let's look at them separately.

Sensory dissonance

We don't fully understand how sensory dissonance works yet, but we do know that unlike musical dissonance, it's not a matter of subjective opinion; it's a feature of how your hearing system and your brain register sound. Neuroscientist Anne Blood and her colleagues at McGill University have shown that activity in an area of the brain that is associated with unpleasant emotions (the parahippocampal gyrus) increases with increasing levels of sensory dissonance.[1] Basically, sensory dissonance is a biological effect that makes us experience certain combinations of notes as unpleasant. From the day we are born (and perhaps earlier), we have a general preference for consonant (non-dissonant) combinations of notes.[2] Even baby chicks have been found to prefer consonance to dissonance.[3]

Although we can learn to enjoy a certain amount of sensory

dissonance because it adds spice to music, we don't lose the ability to distinguish between consonance and dissonance. People become indifferent to sensory dissonance only if the appropriate part of the brain (the parahippocampal cortex) is damaged.[4]

When your eardrum vibrates in response to a musical note, it passes the vibration on to a part of your inner ear called the cochlea. This cunning device sorts out the notes in the same way that a shop assistant sorts out the jackets on a clothes rack. Jackets are placed in order of ascending size—and the notes are organized in order of increasing frequency. The fine details of how this is achieved are, to put it bluntly, completely beyond me, but thankfully we only need to know the basics.

The cochlea is a small, coiled tube, and vibrations passed on from the eardrum enter it at one end. If you listen to a single note with a certain frequency, the vibrations will travel into the tube, causing excitement among some of the tiny hairs on its inner surface. Each frequency of vibration is linked to a particular place somewhere along the length of the tube. The note you hear will have not much effect on the tube in general but will make the hairs dance like crazy in a particular area. The dancing hairs then send off a message to the brain saying, "Our frequency has arrived—hurrah!" The brain recognizes which group of hairs is involved and gets the message that we are listening to a note of a certain frequency. Very-high-frequency notes find their dancing hairs just inside the entrance to the tube. Bass notes find theirs down toward the other end—as you can see in this sketch:

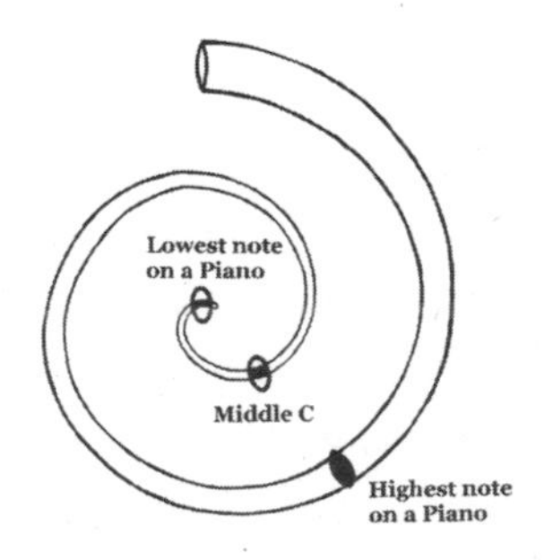

This all works very well unless the notes are too close together in frequency. Each frequency excites a little patch of hairs, and if the patches are clearly separated, then the brain can make sense of it all. If, however, the patches overlap a bit, the brain becomes confused and unhappy.

One way of demonstrating this confusion and unhappiness is to use another of your senses—your eyesight.

Have a look at this:

NNOOTTEE

This image is made up of two overlapping words and it's very difficult to recognize which words are involved. If the words are printed side by side, though, without any overlap, they become easy to read:

NOTE NOTE

And if they are printed with almost complete overlap, it's fairly easy to work out what the print says:

NOTE

So, somewhere between almost complete overlap and no overlap at all, your vision system can be confused and made to work hard trying to untangle what it's looking at.

The same sort of thing happens with your hearing system. If two simultaneous notes are very close together in frequency, the brain recognizes only one (slightly expanded) patch of dancing hairs and therefore one note. If the patches are clearly separated, then the brain recognizes two notes. Somewhere between these two extremes the combination of partially overlapping frequencies sounds dissonant, just as the partially overlapped words we saw earlier were the most difficult to read.

This is why, if you press down any key near the middle of a piano keyboard and the key next to it, you hear a jarring, dissonant sound. Notes a semitone apart played together make for a rough, tense combination wherever you are on a piano keyboard because their frequencies partially overlap on the cochlea.

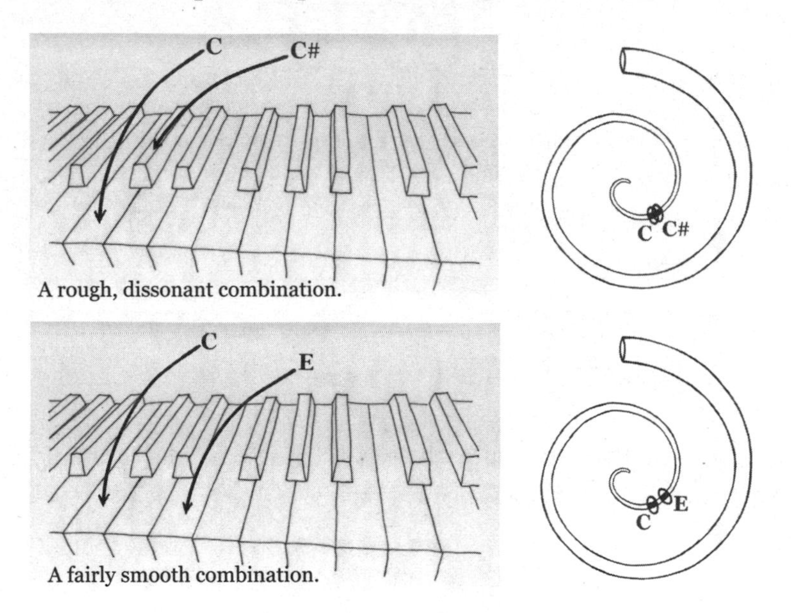

A rough, dissonant combination.

A fairly smooth combination.

If you play notes two semitones apart, the notes still sound fairly dissonant. When the distance between the notes is three or four semitones, everything starts to sound more pleasant and relaxed. But that's true only on the right-hand half of the piano, where the mid-range to high notes are.

Down on the left-hand, bass half of the keyboard, things get tricky. The patches of dancing hairs that respond to low notes are closer together than the ones for high notes—so nearby notes are more likely to overlap. As we move downwards in pitch, notes need to be farther and farther apart before they sound agreeable. Down in the deep bass section of the piano, even normally harmonious, large intervals sound rather dissonant. (I use the term "rather dissonant" here because sensory dissonance isn't an on or off thing; it's more like a sliding scale from "very dissonant" to "consonant.")

Down at the bass end of a keyboard, notes a surprisingly long way apart overlap on the cochlea. Even the combination shown here (seven semitones, where the upper note has a cycle time that's ⅔ of the cycle time of the lower note) will be rather dissonant.

Composers, songwriters, and arrangers may not know the scientific details of why harmonies sound poor for bass instruments if the notes are anywhere near each other, but they can hear that it happens, which is why they write chords and harmonies with big gaps between the notes for bass instruments. This phenomenon also explains why chords that sound clear and harmonious on a standard guitar sound far less so on a bass guitar: your brain can't distinguish the low notes from one another clearly, so everything sounds woolly and dissonant—which is why most bass players don't play chords.

As I said earlier, we don't fully understand sensory dissonance yet, and although this "overlap on the cochlea" theory provides a good general description of what's going on, scientists are still busy at work on the subject.

But let's not forget that sensory dissonance isn't necessarily a bad thing in music. If we add a dissonant note to a chord, we get a tense sound that needs to be relaxed sometime soon. A lot of music uses this effect to build a cycle of tension and release that adds motion and interest to a piece. Without the occasional use of dissonance, music would be bland and far less effective at stirring our emotions. So in spite of the fact that we begin life with an aversion to sensory dissonance, we can grow to appreciate and enjoy it.

Musical dissonance

Rather than being based on biology (like sensory dissonance), musical dissonance is generally a matter of taste. How edgy do you like your harmonies or melodies or rhythms to be? Do you like everything to be ordered, relaxed, and musically consonant, or do you like your music to be spicy, risky, and musically dissonant? How much do you enjoy unfamiliar types of music?

Musical dissonance sometimes includes some sensory dissonance (e.g., spiky jazz chords involving notes a semitone apart), but that's not necessarily the case. For example, if a songwriter or composer writes something soothing in the key of B major and suddenly interjects an F major chord, the event is musically dissonant*—even though the F major chord itself is consonant. The chord hasn't caused any confusion in your cochlea. It's just that you weren't expecting it.

Basically, we call a piece of music musically dissonant if it's outside our comfort zone.

* None of the notes that make up an F major chord are members of the key of B major.

CHAPTER 14

How Musicians Push Our Emotional Buttons

Whether they are performing a written piece or improvising, musicians have an array of techniques they can use to maximize the emotional effect of the music. In this chapter we are going to look at musical performance over a wide range of styles to see how musicians cast their spells over us.

Performing musicians in different genres have differing levels of control over what you hear. A classical keyboard player performing a Bach fugue on a harpsichord must play all the notes in the right order and has control only over the speed of her playing. She can slow down and speed up at various points and introduce dramatic pauses, but that's about all the flexibility she has. When the same musician is playing a romantic piano piece by Chopin later in the week, she has slightly more influence over the sound because on a piano you can vary the loudness of the music. But once again, all the notes must be produced in the correct order.

On Saturdays our busy musician plays in a pub rock and blues band. When they're performing Led Zeppelin's iconic "Stairway to Heaven," they know that they need to be as faithful as possible to the original recording (because a lot of people are familiar with it), but no one is going to complain about a few missing harmonies or a simplified guitar solo. By midnight the band members are meandering their way through a twenty-seven-minute version of

"The House of the Rising Sun." Audience expectations are completely different in this case. It's a traditional folk song and the famous 1964 cover of it by the Animals is a well-known launchpad for endless blues improvisation—so, apart from the last two minutes and the first two minutes, the band will be improvising and free to produce whatever notes they want, as long as they stay within the standard rules of blues improvisation.

In all these musical environments our keyboard player will be using various techniques to give the music its maximum emotional impact. The same is of course true of musicians from traditional non-Western genres, who improvise entire performances without reference to songs or tunes the audience might have heard before.

We'll start our discussion of these techniques by looking at Western classical musicians who are trying to reproduce a piece of music written as dots on a page several hundred years ago. How do you inject any human emotion into a performance like that?

It is, as I have said in earlier chapters, possible for a computer to evoke an emotional response in a listener simply by producing the right notes in the right order, but real musicians do the job much more effectively. As the authors of *Psychology for Musicians* put it:

> The crux of expressive performance is nuance. Nuance is the subtle, sometimes almost imperceptible, manipulation of sound parameters; attack, timing, pitch, loudness and timbre, that makes music sound alive and human rather than dead and mechanical.[1]

Phrasing

One of the main differences between a dry performance by a computer and a more emotional one by a musician is that the musician will automatically (and usually unconsciously) divide the music up into phrases. Phrasing is as important to music as it is to speech because it allows the listener to experience the music as bite-sized chunks rather than a continuous stream of notes. The core reason behind our preference for phrased musical sounds is linked to our predilection for grouping items or events together,[2] which we covered in chapter eleven.

In some cases the beginning and end of a phrase is obvious. The tune to the line "Baa, baa, black sheep, have you any wool?" is a musical phrase that clearly begins on the first "Baa" and ends on the long note on "wool" (tunes often end on long notes). But lots of pieces of music consist of an unbroken stream of equal-length notes, and the beginning and end of each phrase is not at all clear if you just look at the printed score. If a simple computer program performs this sort of piece, it will produce a sequence of equal-length, equal-loudness notes, but we listeners will automatically use our previous musical experience to divide the music up into phrases that are usually only a few seconds long. Human musicians will make things a lot easier by phrasing the music for us, but what are they actually doing?

Modern recording techniques have made it possible for musicologists to make a detailed study of the loudness and duration of every note in a performance. This type of analysis has revealed the fact that musicians use subtle (and sometimes not so subtle) variations in loudness and speed during each phrase. In general, the middle part of each phrase tends to be slightly louder and faster than the beginning or the end.* Because most phrases last for only a few seconds the performer is constantly varying the volume and speed, but the changes are usually so subtle that we don't consciously notice them.

Obviously, musicians are also listeners, and they pick up the concept of phrasing from all the music they have heard and then natu-

* As usual I'm generalizing here. There are lots of exceptions to the observational "rules" I present in this chapter.

rally apply it to the music they produce, in the same way that we learn to phrase our spoken sentences from listening to others. But some of our musical phrasing might be instinctive rather than learned. Psychologists Amandine Penel and Carolyn Drake have looked into the possibility that slowing down at the end of phrases is partly involuntary.[3] They suggest that our hearing system tends to make the final two notes of any group sound closer together than they actually are, so musicians automatically stretch out the timing at this point to compensate for the perceived rush. They recorded eight professional pianists playing a romantic piece in a normal, musical way and then asked them to play the same piece with no emphasis, as a machine would. On average the musicians, when they were playing normally, slowed down by about 30 percent toward the end of a phrase. Later, when they thought they were playing like an accurate robot, they were in fact still slowing down at the same places, but only by about 10 percent. The musicians couldn't hear themselves slowing down; it just happened automatically.

Another piece of research into phrasing by a team led by Finnish psychologist Eva Istok involved listening to a series of four recordings with a range of phrasing styles:

1. extreme phrasing (unusually high levels of acceleration toward the middle of a phrase and equally unusual slowing down at the end)
2. normal "expressive" levels
3. deadpan (no phrasing)
4. inverted phrasing (slower in the middle of phrases and accelerating toward the end)

They found that melodies played with inverted phrasing were rated as sounding less pleasant than any of the other phrasing strategies, including the deadpan performance. I've never heard any music with inverted phrasing because it's extremely counterintuitive and you need a computer program to manipulate the speed of the notes, but I'm sure Eva and her colleagues are right when they say that "inverted phrasing simply sounds wrong."[4]

Emphasis

As their training develops, musicians learn that some types of music benefit from more exaggerated phrasing than others. A romantic prelude by Rachmaninoff can survive overdose levels of phrasing, but a piece by Bach can sound forced or amateurish if the phrases are emphasized too heavily.

Whatever the type of music involved, we can think of each phrase as having its own arc of loudness and speed: increasing toward a maximum somewhere near the middle, and diminishing again at the end. Pieces of music generally last several minutes, however, and if the musician just kept on repeating the same loudness/speed arc for each phrase, the piece as a whole would sound lifeless. Most pieces of music have a main climax and several minor climaxes on the way, and the musician needs to build the music up to these climax points and then relax it after them to give the whole experience an ebb and flow of emphasis. During the course of the piece a sort of architecture is built up of arches within arches.[5] Like this:

Written music for a solo performer is like a written speech, and the musicians need to do the same job as a good orator or actor: they have to breathe life into it all by using a hierarchy of emphasis. Subtle differences in emphasis can have a big effect on the emotional content of a piece of music just as they can in speech.[6] For example, if I emphasize different words in the simple sentence "Fred loves Pete's banjo recordings," we get a variety of meanings. In each of these examples I've put the emphasized word in italics, and the statements in parentheses show the alterations in meaning:

Fred loves Pete's banjo recordings. (But no one else does.)
Fred *loves* Pete's banjo recordings. (He's surprisingly keen on them.)
Fred loves *Pete's* banjo recordings. (But he doesn't like recordings by other banjoists.)
Fred loves Pete's *banjo* recordings. (But dislikes recordings of Pete playing the flute.)
Fred loves Pete's banjo *recordings.* (But doesn't like Pete's live performances.)

Musicians can use a number of techniques to emphasize the different levels of climax within a piece of music. As we've just seen, loudness and speed are very important tools, but deliberate timing adjustments, dramatic pauses, and changes in timbre can also have a big effect.

Rubato

Classical musicians are trained to obey the composer's instructions, which are simply represented as dots on a page together with a few, often vague written instructions such as "almost walking speed" or "majestically." (Traditionally these instructions are written in German, French, or Italian: *andantino,** *maestoso.*)

---

* Just to make life extra difficult, "andantino" has two meanings: slightly faster *or* slightly slower than walking speed. You just have to guess which one the composer had in mind.

One instruction that occasionally crops up is the Italian word *rubato,* which means "stolen." If this word occurs in the score, it means that the musician should play some notes and passages quicker than they are written, and others should be played more slowly, to bring out the maximum emotional effect. Composers use the word "stolen" because some of the time available for the whole piece has been stolen from certain sequences of notes and given to others. This effect is usually associated with romantic music like pop ballads and composers such as Rachmaninoff and Chopin, rather than music that requires a steadier beat like techno or Bach.* Musicians use their judgment in choosing where to hurry and where to dawdle, and the most common technique is simply to make the short notes a bit shorter and long notes a bit longer.

If the score doesn't include the instruction "rubato," the written notes have, in theory, precise pitches and lengths, but the musicians, particularly the experts, don't obey the instructions exactly. Expert solo musicians often take far more liberties with the written rhythm than competent amateurs do.[7] But of course there is nothing arbitrary about these timing changes: the experts are using rhythm adjustments in an attempt to maximize the emotional effect of the piece. Careful observations of virtuoso classical pianists have revealed that, although they make considerable changes to the lengths of the notes involved, they do so repeatably. Asked to play the same piece months later, they play the work with almost identical stretching and shortening of particular notes and phrases unless they have specifically decided to rethink how they play the piece. And this ability to reproduce the nuances of a performance doesn't just apply to classical musicians. Musicologist

---

* Dance music needs a repetitive beat so you can coordinate your movements to it. J. S. Bach and some other composers often build up elaborate interweaving patterns that require the clarity of a steady beat.

Richard Ashley has found this type of reliability in separate recordings of "Yesterday" by Paul McCartney.[8] You might assume that this type of performance stability comes from memorizing individual note values and emphases, but that's not the case. Instead, the performer is only remembering a general performance strategy for that particular piece which has worked in the past.

Deliberate mis-timing

Lots of pop singers add extra emotion, life, and movement to their performance by singing occasional notes a little early or a little late. I was going to spend hours listening for a good, clear example of this, but fortunately Howard Goodall, author of *The Story of Music,* has already found one for us.[9] Here's the first verse of Bob Dylan's song "Make You Feel My Love":

When the rain is blowing in your face
And the whole world is on your case
I could offer you a warm embrace
To make you feel my love

If you listen to Adele's 2008 cover of the song, you can hear that she sings the last two notes of the first three lines early. Listen for the words "your face," "your case," and "embrace." If you tap along with a regular rhythm, you'll hear that these words begin just before your tap. The written music would place the beginning of each word on a tap—but in fact your tap will hit the middle of the word instead. Singing words early is very effective and it's also quite difficult to do. I've been trying for almost ten minutes now and can't get anywhere near it. So there's no reason to feel anxious, Adele. I won't be applying for your job anytime soon.

Loudness and timbre

Even if a piece is played with inhuman precision by a computer, there are parts of the music that are more significant than others—the culmination of a climb up to a long, high note, for example. One way for a (human) musician to add extra emphasis at these moments is to pause just before the climax, or to play the climax unexpectedly loudly, or even quietly. It's the unexpected nature of the emphasis technique that matters, so a sudden reduction in loudness can be just as effective as an abrupt increase.

Changes in timbre can also be used as an emotion-enhancing tool. (The timbre of an instrument is its distinctive sound or voice.* It's why a saxophone sounds different from a banjo even if they are playing the same note.) The timbre of some instruments can be quickly changed from smooth to harsh and back during the course of a piece. This is another way to emphasize climaxes or to create emotional lulls within the music.

And I think this is the point at which we must take a few minutes out to feel sorry for harpsichord players...

Harpsichords were the precursors to pianos, but for reasons that will shortly become apparent, they fell rapidly out of favor as soon as the early piano designers finally got pianos working properly. The British conductor Sir Thomas Beecham described the harpsichord as sounding like "two skeletons copulating on a tin roof in a thunderstorm," which is a bit harsh, although it is rather jangly and clunky-sounding. But the big problem with harpsichords isn't their jangly sound; it's the fact that the player has no control over the loudness or timbre of the notes.

A note on a harpsichord is produced by pressing down on a piano-like key, but that's where the similarity between the two

* For more information on timbre, see "Fiddly Details," section A.

instruments ends. On a piano, the key you press is attached to a series of ingenious levers that eventually fling a felt-covered wooden hammer at strings (there are two or three strings for each note). The quicker you depress the key, the faster the hammer is traveling when it hits the strings. Quicker hammers give louder notes with a slightly harsher timbre—so the pianist can control the level of emphasis.

In the case of a harpsichord, the pressing of a key moves a lever that has a small spike (made from the quill of a crow's feather) sticking out of it. When the key is pressed, the spike plucks the string to produce a note. And no matter how fast or hard you hit the key, the plucking action doesn't change; the string can't be plucked harder or softer to produce a louder or quieter note. So on a harpsichord, neither the volume nor the timbre of the notes produced can be altered,* which means the only emphasis control the player has is to speed up or slow down. This limitation is the prime reason why the solo harpsichord has a very low tear-jerk rating and is rarely used as an accompaniment to kissing scenes in films... unless the couple doing the snogging are vampires.

Pianos, as I just mentioned, do offer control of the loudness, but although the timbre changes slightly, it is not actually controlled. All that happens is that an increase in loudness is accompanied by a slight increase in harshness of tone because the louder note contains a slightly greater proportion of its higher harmonic frequencies, which makes the sound a little more aggressive.

Timbre control is better on stringed instruments because, as you pluck or bow the string closer to its end (i.e., nearer the bridge), the timbre gets more metallic and tense-sounding.† Saxophones and other wind instruments are also pretty good for timbre

* Some of the bigger harpsichords have two keyboards that have different sounds, to make things more interesting, but the basic problem is still there.

† See "Fiddly Details," section A, for further details.

manipulation, but arguably the best instrument for timbre control is the human voice. For the voice, as for other instruments, our perception of different timbres having different emotional content is probably based on our experience of hearing people talk in different emotional situations. When we are calm, our vocal cords are relaxed and we don't use much breath to speak. The passage of low-pressure air over relaxed vocal cords means that the "notes" we produce as we speak contain a very low proportion of high harmonic frequencies. But if we get angry or frightened, our vocal cords get tighter and we use more air to speak. This combination introduces a greater proportion of higher harmonics into the vocal mix, and we therefore associate this type of timbre, even in musical contexts, with tension, harshness, and excitement.[10]

Legato and staccato

On top of all these variables there is also the choice between producing the notes as a sequence of separate entities (as you might say the words "What do you mean?") or running the sounds together in a joined up slur ("Whaddyamean?"). The "running notes together" technique is called *legato,* and the "distinct separation of notes" is called *staccato.* Legato is more common than staccato and generally doesn't just involve making the notes join end to end in a continuous stream; there's often an overlap between them. On many instruments it's possible to let the previous note keep on sounding for a while after the next note has started. This is a very common technique in guitar playing and a central feature of harp music. Pianos can do it, and if the right-hand pedal on a piano is pressed down, the overlaps between notes become much more pronounced. This pedal is often mistakenly called the loudness pedal, but it doesn't make anything louder. It just allows the notes to ring on for a while so the sounds overlap.

The king of note overlap is the organ. On pianos, harps, and guitars, the earlier note is always dying away as the subsequent notes are played, but on an organ, the old notes retain their power as the new notes are introduced. You can start off with one note and add others one at a time until you've got a huge powerful chord with up to ten notes in it. J. S. Bach uses this technique a few seconds into his Toccata and Fugue in D Minor just after the famous opening flourish.

It's impossible to produce overlapping notes on a wind instrument such as a saxophone or flute because they are designed to produce only one note at a time. But wind instrument players can at least make the notes join end to end by producing a stream of notes in one breath. For the "separate note" staccato effect, a wind instrument player will provide a separate burst of air for each note, and a piano player who wants to play staccato will use the left-hand pedal, which damps the sound, making the notes die away more quickly than usual.

Playing music staccato tends to increase the tension and energy in the sound, and legato makes everything sound smoother and more relaxed.

So there we have it. There are lots of ways of adding emotional content to a musical performance—even if the starting point is a couple of thousand dots on a page written down three hundred years ago.

Of course all the techniques I've discussed so far are available to musicians across all genres, and they use them all the time—whether the music is composed by a jazz musician, written by a garage band, or improvised by a sitar player. Nearly all musicians use phrasing, loudness, and speed variations, emphasis, timbre, and detached or slurred-together notes. But there are some exceptions to this free-for-all. While many musicians use rubato to some extent, there is one group of musicians who never do: African drum ensembles. One of the oldest African musical genres is drums-only music, which often employs polyrhythms. A polyrhythm is produced when two or more drummers follow different but linked agendas at the same time, creating complex and interesting effects from a combination of rhythms. For example, if one drummer taps with a completely steady beat and emphasizes every fourth beat, you would hear the equivalent of this pattern:

I I I **I** I I I **I** I I I **I** I I I **I** I I I **I** I I I **I**

Boring, eh?

Now, if you ask that drummer to stop playing and get a second drummer to tap along at the same speed, emphasizing every third beat this time, you get another uninspiring rhythm:

I I **I** I I **I** I I **I** I I **I** I I **I** I I **I** I I **I** I I **I**

But if you ask both drummers to play their dull rhythms at the same time, you get something quite complicated:

I I **I** **I** I **I** I **I** **I** I I **I** I I **I** **I** I **I** I **I** **I** I I **I**

(The biggest I's show when the two drummers both emphasize a beat at the same time.)

With this simplest of polyrhythm rules (try it with a friend, tapping on a table), we have created an emphasis pattern that repeats only every twelfth tap and would be considered very jazzy and advanced in Western music. It would, for example, fit very well into a complicated classical piece like Stravinsky's *Rite of Spring*. But to a couple of African drummers, this rhythm would be laughably simple. Members of an African percussion group who had never heard Western music before might find tunes and harmonies by Beethoven or the Beatles interesting, but they would consider the rhythms simplistic to the point of being childish. Using polyrhythms, they might be working in patterns of sevens and nines with double and triple taps rather than the single ones in our simple example. Because polyrhythmic music requires a strict underlying beat, rubato must be avoided because it would muddy the effect of the drumming patterns coming in and out of synch with one another.

And I see we've drifted into the realm of world music—which brings us nicely into a discussion about improvisation.

Improvisation is simply making music up as you go along—and my use of the word "simply" here is completely misleading, because there's nothing simple about it.

Western classical music composers writing between about 1800 and 1900 frowned on improvisation as the unworthy pastime of ne'er-do-wells. That attitude has, in most cases, persisted in the genre to this day, and apart from minor deviations in timing for emotional effect, Western classical musicians are generally taught to follow the instructions of (usually dead) composers precisely. Most other Western music genres also concentrate on faithful re-creations of composed pieces, with only a smattering of improvisation.

As far as most of the rest of the world's traditional musicians are concerned, this approach is deeply weird. Professional musicians from non-Western traditional musical cultures train just as hard as Western professionals, but much more of their training concentrates on the ability to improvise.

At this point I need to explain a bit more about improvisation. When you improvise, you don't allow yourself total freedom. You choose your notes carefully, according to the rules appropriate to whatever genre of music you are playing—and there are always rules. Rule-free improvisation would be like rule-free conversation: a frustrating, unrewarding mess. This comparison with conversation is quite useful because conversations *are* improvised. You piece together your conversation by following a theme and expanding upon it while carefully monitoring what your friend is saying. You are (subconsciously) juggling dozens of rules and guidelines that prevent you from making mistakes. For example, if your friend says her new haircut makes her look stupid, do *not* enthusiastically agree. Also, we don't keep inventing entirely new sentences. We use stock phrases such as "I'm looking forward to it" and "See you later." Improvising musicians often use standard musical phrases in exactly the same way.

In Western music, jazz is the most improvisatory genre. Many jazz fans assume that there is almost unlimited freedom in modern jazz improvisation—but the musicians know otherwise. It's true that some jazz ensembles strenuously avoid straightforward tunes, like the mythical band led by saxophonist Progress Hornsby, alter ego of TV comedian Sid Caesar:

> We've got a nine piece band where the ninth member plays radar—to let us know if we get too close to a melody.[11]

But usually, beneath all the seeming spontaneity and complexity, there is a musical agenda understood by all the musicians whether there is a tune involved or not. Here's what the world-renowned jazz pianist Dave Brubeck had to say about the rules of improvisation in jazz circles:

> Interviewer: Are there any rules for improvisation?
> Dave Brubeck: You bet your life there are, and the rules in jazz would just scare you to death. They're so strict it's fearful. Just break one of the rules and you'll never end up in another jam session with the same guys again.[12]

We'll come back to modern Western music later, but I want to return to what all those non-Western musicians were doing. How do they train to improvise?

Let's take the example of Iranian classical music ("classical" here meaning long-established music that requires a lot of training). To become a full-fledged Iranian classical musician, you must first memorize the Radif, about three hundred short pieces of music, each between thirty seconds and four minutes long.[13] These short tunes aren't considered complete pieces in their own right; they are the building blocks from which you will produce your improvisations. It takes about four years of hard study and practice to learn your three hundred pieces, and it would take eight to ten hours to perform them all end to end. But of course they are not supposed to be played that way. The idea is that a performance will be built up of variations on a few of the pieces.

In Arabic, Turkish, or Indian music the rules are rather looser. As a musician starts a piece, there are four basic ingredients involved:

1. An idea of the mood the musician is trying to produce.
2. The chosen selection of notes he or she will be using.
3. A selection of short tunes that are typical for the chosen mood and note combination.
4. Many thousands of hours of practice and training.

One of the most common forms of improvised Indian classical music is the raga, often performed by a sitar player together with a percussionist and a third musician playing a tanpura, an instrument that provides a continuous drone. Traditionally ragas are in four sections. For the first few minutes the sitar player plays slowly without a clear rhythm. To Western ears this sounds like he is just warming up. But what he is doing is introducing the audience to the family of notes he will be using during the piece.

This family of notes is actually the raga. A raga is in some ways similar to a Western key in that it generally has between five and seven notes in it. But a raga is more than just a collection of notes. Each raga comes with a set of rules governing how the notes are to be used in certain situations. Some ragas might use only five of the notes on the way up but seven on the way down. Others have a rule that, having climbed to a certain note, you must double back on yourself a note or two before you can continue upwards (and the same type of rule might apply to a different note on the way down). It's also common for certain notes to be emphasized more than others. As a result of such rules, different ragas have individual "catchphrases" that help the audience identify them.[14] Identifying the raga can be difficult for the untutored Western ear, partly because sitar players bend the strings of their instrument even more often than blues guitarists do, and notes are rarely played without some embellishment or other. But of course you don't have to know anything about the raga to be able to enjoy the music.

Once the slow, meandering introduction is finished and the group of notes has become familiar to the audience, the sitar player starts to give the music a definite pulse for the second section. The improvisation then often gets fast and furious in the third section before, finally, the percussionist joins in for the finale.

The rhythms used in ragas are also subject to strict rules. The rhythms are not generally written down but are learned with the help of mnemonics such as "TakitaTakitaTaka," which, if you say it without gaps (emphasizing the three "Ta" syllables), gives you a 1,2,3,1,2,3,1,2 rhythm. (It's best to repeat the mnemonic two or three times to hear the rhythm properly.) One of the most important rules of raga improvisation is that you must never lose your place in the rhythm, which can be very tricky when you are dealing with complex rhythmic patterns. Whereas Western rhythms can be identified after half a dozen notes, one of the Indian rhythms takes 108 beats to complete its full cycle.[15] The percussionist (who usually plays the tabla, a pair of hand drums of different sizes) and the sitar player often take turns with short rhythmic phrases, chasing each other around at incredible speeds—but their years of training ensure that they are always completely in control, synchronizing exactly whenever they want to, particularly on the final note of any climax.

Improvisation in modern Western music

Jazz from the 1930s onwards and rock in the 1960s and 1970s often involved improvised solos and duets in the middle of performances of prewritten pieces. The general idea was that the harmony of the original piece would continue and the soloists would devise new melodies over it. The new tunes would follow the standard melody rules we saw in chapter ten, for example:

- Lots of small jumps in pitch with few large jumps.
- Big jumps upwards are usually followed by a smaller jump downwards.
- Emphasis of notes when the tune changes direction.
- Use of notes **1** and **5** at points of rhythmic emphasis.

And one extra rule that we haven't come across so far:

- The emphasized notes in tunes will usually be found in the harmony that is playing at that moment. (So, for example, a rock, jazz, or blues band will often be going through a predictable cycle of chords while the lead guitarist performs a solo. If the chord at any point includes the notes A, C, and E, then the guitarist will tend to use one or more of these three notes as the emphasized notes in his solo. As the chord changes to another group of notes, one or more of these new notes will now be emphasized. Written or composed music also tends to follow this rule, but in that case the chords are chosen to follow the emphasized notes in the tune, rather than the improvised tune following the chords. In both cases the chords and the tune usually have some notes in common.)*

Obviously the musician wouldn't be running through these rules consciously; he may not even know about them. The whole process is automatic and a consequence of all those hours of practice. This type of improvisation is still a feature of many rock and blues concerts: you play half a song, then improvise over the song's harmony, and finally you finish off with the second half of the song.

Although some jazz musicians still follow this tried-and-tested recipe, the innovators have experimented with different approaches. Bebop, the new jazz of the 1940s, was an extension of the standard technique and involved chords that had five or six or more different notes in them instead of the standard three or four. This gave the soloists more freedom to choose whatever notes they emphasized in their melodies—but the improvisation was still based on the harmony. In the late 1950s, trumpeter Miles Davis started to experiment with the idea of melodic improvisations that involved choosing a certain collection of notes, to be played over simpler chords that didn't change much (a similar concept to the Indian raga). This technique, called modal jazz, is the basis for the most popular jazz album ever produced—*Kind of Blue*.

An improvising member of a band is involved in a musical conversation with the other band members, and this involves a lot of fast thinking. In real conversations you will no doubt be familiar with the irritating phenomenon of thinking of *exactly* the appro-

* For more about how this works, see "Fiddly Details," section C, "Harmonizing a Tune."

priate response to a comment approximately fifteen minutes too late. But we can't all be Winston Churchill:

> Lady Astor: Winston, if you were my husband I'd poison your tea.
> Churchill: Madam, if you were my wife I'd drink it.

During any improvisation the musicians (like Churchill) have to make split-second decisions about how to respond to what's going on. They'll often play it safe and stick to note sequences they know well, which explains why so many 1970s lead guitar solos have similar bits in them. I suppose someone must have been the first to play diddle-doddle-diddle... diddle-doddle-diddle... diddle-twang on the electric guitar... but his name was lost in the clouds of smoke from patchouli-flavored joss sticks back in 1968. Individuals have their own favorite building blocks from which to construct solos (rather like the Radif mini-tunes I mentioned earlier). You might think that your favorite guitarist is far too good an improviser to be relying on a collection of favorite clichés, but the reason you can recognize a solo by him within a few seconds, even if you've never heard it before, is that he has a distinctive improvising style, and distinctive styles depend upon a collection of personal clichés.

But you're less likely to chance upon a seam of beauty if you just stick to the obvious choices and clichés, so highly skilled musicians occasionally opt for more unusual notes or chords. Sometimes this pays off and something sublime happens as two or three musicians suddenly shift direction together. More often than not it just sounds OK. And, of course, sometimes the adventurous choice turns out to be something everyone regrets.

Although some improvisation mistakes are the result of an error in judgment, it's far more common that a wrong note is simply the result of misplacing a finger on the instrument or blowing too hard. The important thing, as far as the musician is concerned, is what to do after you've played a wrong note. How do you recover? The answer is as simple as it is surprising: the best thing to do is to repeat the error. Play the same passage again and make sure you include the wrong note. Doing this has two effects. First of all, because the unexpected note has just been heard, when it comes around again it sounds less unexpected and fits in better.[16] Second, repeating your mistake misleadingly confirms to the listener that you intended to play that note in the first place. You're basically trying to pull off a con trick by saying to your audience, "Oh, no, I didn't make a mistake. I did exactly what I intended to do. It's not my fault if you are too unsophisticated to appreciate my adventurous choice. Anyway, I'm prepared to forget the whole misunderstanding if you are."

This is quite a complicated message to get across in one bum note—but it often works.

When improvising musicians aren't making mistakes and callously blaming the listener for them, they often use the characteristics of the human voice to create emotional responses. I mentioned earlier the effective technique of making the timbre harsher during the emotional climaxes because our voices do the same thing, but improvising musicians can take this one step further because they can also choose the notes being played. Using computer-generated note sequences, psychologists Klaus Scherer and James Oshinsky have demonstrated that we associate high pitch levels, rising pitch contours, and a fast pace with anger and fear, probably because angry, fearful, or excited people talk quickly at a high pitch that rises toward the end of statements.[17] Improvising musicians tend to use all these features when they want to reach an exciting climax. If you listen to any recordings of improvised music, you'll find very often that emotional climaxes involve harsher timbres, increased loudness, increased pitch levels, and rising pitch contours.

Whatever the genre, from raga to reggae, composed or improvised, we'll occasionally go to an exceptional concert and hear a truly great, deeply moving performance. At these times it's easy to think that the greatness is an intuitive and effortless outpouring of superhuman musicianship, but the unglamorous truth is that great performances are the result of a lot of repetitious, detailed hard work.[18]

Since the 1980s there has been an increasing trend toward collaboration between musicians from different genres and cultures. In the UK the WOMAD (World of Music, Arts and Dance) festivals, which began in 1982 under the leadership of Peter Gabriel, have grown in strength, and what used to be a minor interest has become the vast genre known as world music.* As far as music is concerned, we've never had it so good.

---

* If you want a great introduction to world music, I recommend *Real World 25*, which is available as a three-CD boxed set or MP3 download. You'll find a list of suggestions for listening and watching in all sorts of genres at the back of the book—but this is the one I'd recommend most of all.

CHAPTER 15
Why You Love Music

Music, particularly singing, is an important feature of all human societies, and we have evidence that it has been so for many thousands of years. In 2008 a team of archaeologists working in southwest Germany discovered a selection of flutes that are approximately forty thousand years old.[1] These instruments were made from vulture bones with finger holes drilled in them and they had been very carefully crafted. Whoever drilled the holes had first cut fine scratches in the bone to show exactly where the holes should go. Wulf Hein, a member of the archaeological team, made a copy of one of these flutes and found that it produces a pentatonic scale, the five-note scale that has been the basis of most musical systems throughout history, including the Western major scale. Many tunes are based on the pentatonic scale, including "The Star-Spangled Banner," which is what this amiable German researcher performs on his replica flute on YouTube (search for "the world's oldest instrument" if you'd like to hear it for yourself).

Of course, we have no direct evidence that prehistoric humans did a lot of singing, but we can be confident that they did so from anthropological studies of modern hunter-gatherers. Sometime between 1 and 2 million years ago, humans had evolved to the point where they had much bigger brains than other primates and lived in hunter-gatherer societies. Agriculture was developed only about ten thousand years ago, which means that for more than 99 percent of human history, all humans were hunter-gatherers.[2] This way of life has almost died out nowadays, but back in the latter half of the twentieth century, there were still enough hunter-gatherer tribes around in Africa, Asia, Australia, and South America to keep hordes of anthropologists in gainful employment. These academics discovered that the lifestyle of the majority of hunter-gatherers hadn't changed for thousands of years. In general, they lived in bands of between twenty and fifty people and moved from place to place, hunting animals and gathering nuts, seeds, berries, and vegetables.

The anthropologists also found that, although the hunter-gatherer lifestyle can sometimes be hard, the working hours are surprisingly good. Studies of several of these societies have concluded that their members typically spend only twenty hours a week hunting and gathering and another ten to twenty hours preparing food and doing other types of work (making or mending tools, etc.).[3] A thirty-hour working week means that they have quite a lot of free time. Anthropologist Marjorie Shostak found that the hunter-gatherers she studied spend most of their recreational time "singing and composing songs, playing musical instruments, sewing intricate bead designs, telling stories, playing games, visiting, or just lying around and resting."[4] The fact that the first three items on this list involve music gives us a sense of its importance in this sort of society. Given the unchanging nature of the hunter-gatherer lifestyle, we can safely assume that music has always been a very important component of human life.

One of the basic ideas behind Charles Darwin's theory of evolution is that if a particular activity is very widespread and extremely old, it's probably because that activity is useful to the survival of the species involved. So in the past, music must have had a positive effect on human survival, and this must be one of the main reasons why it exists.

But what use could music have as a survival tool? Darwin

himself puzzled over this question and eventually decided that we must use music in the same way that birds do. His conclusion was that people perform music to show how healthy and skillful they are, in order to attract sexual partners. In fact there is no need to actually perform; you just have to look the part. A group of French researchers led by Nicolas Guéguen conducted a study which showed that a particular young man was twice as successful at getting girls to give out their phone number if he was holding a guitar case.[5]

But if being a musician was the best way to attract mates, every teenage party would be clogged up with spotty hopefuls carrying xylophones and tubas. So there must be more to music than simple sexual display.

There is plenty of evidence that music bonds groups of people together across a wide range of situations, from grief to celebration, and bonded groups are more likely to survive because they cooperate better when things go wrong.[6] Revolutions thrive on songs because they help to draw the revolutionaries into a single group. The same effect can be seen with crowds of sports fans.

Although the words of these group-bonding songs often have some emotional content, that's not always the case. One of the most popular anthems sung by fans of my local soccer team, Notts County, is called "The Wheelbarrow Song." The lyrics to this masterpiece go as follows:

I had a wheelbarrow, the wheel fell off,
I had a wheelbarrow, the wheel fell off.

These moving and uplifting lines are followed by a chant of:

County! *(clap-clap-clap)* County! *(clap-clap-clap)*

There are at least three slightly unconvincing stories about how this song came to see the light of day, but the only reason why several thousand people continue to sing it at every match is that it conveys a simple message—to the singers, and to everyone else: "We're Notts County supporters—and you're not" (with a side order of "and we don't take things too seriously").

Whether you're an ice hockey team, a group of ABBA fans, or members of a religious sect, communal singing will bind you together and encourage you to help one another if things start going badly. Nowadays this bonding might simply mean that you check that all of your group are on the bus before you set off for home. Thousands of years ago it would have led to better cooperation in hunting, or in fighting off enemies or predators. In fact, the bonding that comes from communal singing may well still be saving lives today among platoons of U.S. marines and other soldiers who sing while marching or training. The singing helps the rhythm of the marching, reduces the boredom of the long treks, and helps to make the group into a self-supporting team.

Sexual advertising and group bonding give us part of the answer to the question "Why does music exist?" but the theory I like best concentrates on the importance of mothers* singing to their babies.

Lullabies and play songs

The outstanding feature of human babies, compared to the rest of the animal kingdom, is that they are so totally helpless for such a ridiculously long time. Zebras, for example, have a much more get-up-and-go attitude. A baby zebra can run with the rest of the herd within a few hours of birth and will reach adulthood in about three years. It's true that zebras don't develop to the extent that they can write historical novels or play pool, but a six-month-old

* Not forgetting all you fathers, sisters, brothers, etc.

zebra is a master of predator evasion, while a six-month-old human is just a tasty snack.

Busy parents, from the hunter-gatherers of forty thousand years ago to twenty-first-century advertising executives, have always known that you can't do very much if you are holding a baby. Among apes the answer to this problem is that the baby clings to the mother's body hair, but humans don't have much body hair, so this isn't an option. This means that human babies need to be put down quite a lot of the time, and as we all know, this often upsets the little blighters. Babies are not good at regulating their emotions, and their parents have to spend a lot of time soothing and calming them. Cuddling and stroking are good ways to calm a baby, but if you've just put little Jessica down so you can get on with cooking the evening meal, you need a non-contact method of providing comfort. Singing or humming is an obvious choice. Initially you might try to lull her to sleep with your melodious outpourings, but if this doesn't work, she will still be calmed and reassured by your song, even if she can't necessarily see you.

All human cultures sing to their children and have done so for thousands of years. Plato wrote about the beneficial effects of lullabies on children over two thousand years ago,[7] and the fact that lullabies are similar all over the world is a strong indicator that singing and humming as a means of soothing babies goes right back to the beginnings of human history.[8]

Of course, not all parental singing is aimed at calming the baby down; there are also cheerful and stimulating play songs for youngsters, such as "Pop Goes the Weasel." Babies like it when their mother talks to them in sing-songy motherese, but they love it even more when she sings to them. You might wonder how I can state this point with such confidence. Well, that's because psychologists Takayuki Nakata and Sandra Trehub have carried out an experiment that confirms it.[9] You can tell when a baby is enjoying what her mum is doing because she pays attention. She looks at her mother's face and stops wriggling around so much. Knowing this, our two psychologists played several babies videotapes of their mums either singing or talking and monitored how interested the infants were. The singing clearly won the day.

One probable reason why babies preferred the singing is that songs are more predictable than speech. Mothers usually have a small repertoire of songs that they tend to sing in a ritualized way, so the whole performance is comforting and familiar to the child. In contrast, motherese often involves questions, which babies sometimes find a bit too stimulating. And before you accuse me of overestimating the ability of babies to comprehend questions, I'd better explain myself. If Imogen's mother says, "Who's a good little girl then?" with a motherese lilt, six-month-old Imogen is stimulated, but not because she's considering the nature of goodness and how it pertains to the fact that she's just vomited milk down Mum's silk shirt. Babies haven't got a clue what you're asking, but when you ask a question you usually raise the pitch of your voice toward the end of the sentence. Humans as well as other primates find rising pitches toward the end of utterances stimulating rather than relaxing, since rising pitches indicate that the utterance was important and that a response is necessary. Apart from these frequent questions, motherese is often punctuated by odd noises, hand movements, and tickling. In fact, if you overdo the motherese stimulation, your little one will start averting her eyes, deliberately ignoring you so she can calm down a bit.

In another experiment, Sandra Trehub's team at the University of Toronto monitored the level of cortisol in the saliva of six-month-old babies before and after their mothers sang to them, and they discovered something unexpected.[10] The level of cortisol in your saliva (or blood) is a measure of how stimulated (or stressed) you are. The Toronto scientists expected that the sound of their mothers singing play songs would always lower babies' cortisol levels, and indeed that's what happened if the babies were feeling a little overstimulated and had a high level to begin with. Some of the babies, however, were obviously feeling a bit dopey and chilled out before the singing started and had low levels of cortisol. As Mum started singing, these kids became more stimulated and their cortisol rose to a level that was suitable for paying attention and enjoying the experience.

So play songs are calming and fun for babies if they feel a little stressed, and stimulating and fun if they are feeling dopey. It has also been found that, when children are upset, play songs calm them down more effectively than lullabies.[11] This is because rousing play songs are more engaging and distracting than slower-moving, lower-pitched songs. Lullabies work best when the baby is already feeling content.

Sandra Trehub has also found that North American mothers don't use lullabies as much as mothers in many other cultures. North American mothers interact with their kids face-to-face, singing jolly songs, because their goals are largely playful. In many other cultures mothers spend a lot of time carrying their infants, and there is often less face-to-face contact. These mothers tend to sing lullabies and soothing songs rather than play songs because their goal is to calm their children and/or send them to sleep.[12]

So, as a rough rule of thumb, it's motherese for stimulation, play songs for contentment, and lullabies for sleep.

Clearly, for the first few years of your life one of the most enjoyable things you experience is your mother singing to you (yes, and you, Dad; and you, siblings; and even Great Aunt Geraldine, as long as she puts both of her cigarettes out first).

Maternal singing is a good candidate for the reason why most of us grow up to love music: it's nearly always associated with enjoyment and pleasant outcomes (play or sleep) at a time when we are at our most receptive.

After babyhood we move on to kindergarten, and here, once again, musical activities such as classroom sing-alongs are some of the best parts of our day. The link between happiness and music at this age is very clear: if a child is singing, either to herself or with others, you know she is in a good mood and content with life.

Teenage years

As you reach your teen years, just about every interaction you've had with music has been enjoyable and now it's time to claim your own musical space—one that only you and your friends truly understand. Your musical choices help define you; they are part of your identity.

Historically, the need to build, maintain, and develop a personal identity is a fairly new thing. If you were born in 1730 into a family of farmers, farriers, or fishermen, the basic framework of your identity would be handed to you automatically. You and everyone around you would know roughly who you were and how you were likely to be spending the next few days, months, and decades. Some adventurous folk would create new lives for themselves, but in general, people went with the flow, without realizing that there was a flow carrying them along.

But nowadays it's not so simple.

In the modern world we are surrounded by options about where to go, what to do, and who to be, and the choices we make help to define who we are. For many of us the creation of our musical identity plays a central role in this self-definition process.[13] Music as a tool of self-definition is particularly important during our adolescent years, and we listen to more music during this period than we do at any other time in our lives.[14] In early adolescence our musical tastes tend to be for pop, rock, and dance music. In general we like our music to be defiant (of boring older people and their boring old music), but we tend to align our preferences with those of the main pack of young teens. In later adolescence we become, in our musical taste as well as other things, less randomly defiant but more individual and adventurous. We even gain enough confidence to admit that we sometimes like to listen to Sinatra or the Beatles, despite the fact that most of our peers disagree. Apart from these individual flavorings, though, there is still a general tendency to fall in love with whatever modern subgenre our friends like. As I mentioned back in chapter one, as adolescents we want to feel cool, and optimum coolness comes from having a group of friends who share a definite, but non-mainstream, taste in music. But even if one of the original reasons why you listen to a certain sort of music is to be socially accepted, that doesn't mean that you don't genuinely love it. Music is lovable, and you love the music you are most familiar with.

Adulthood

As we move from late adolescence into early adulthood, we accept a broader range of music into our lives, but unfortunately, most of us don't continue to do so as we get older. American researchers Morris Holbrook and Robert Schindler have looked into this in some detail and have shown that the development of our taste for popular music reaches a peak when we are in our early twenties and declines after that.[15] In their book *The Social and Applied Psychology of Music,* Adrian North and David Hargreaves are clearly amused to tell us that they

> cannot resist pointing out that findings concerning the crystallization of musical taste during late adolescence/early adulthood

might explain the common observation that "Today's music is rubbish compared with that of... [insert year of your choice].[16]

Although we don't seek out new musical experiences with as much enthusiasm as we grow older, we don't actually remain musically frozen in time. We retain a particular love of the popular music we fell for between the ages of fifteen and twenty-five, but we often develop a taste for more complex music, like jazz and classical. This is probably because, by the time we are thirty, we have listened to so much popular music that the genres involved (pop, rock, blues, rap, etc.) have become too predictable and therefore boring. A lot of erstwhile punk rockers and rap fans have no doubt surprised themselves by their slow drift toward jazz and/or classical, but it's simply a sign that they have become expert listeners to straightforward popular music and nothing in it surprises or stimulates them anymore. Finding a new genre that provides enough musical complexity to stimulate this expert brain can be a bit of a struggle at first because it's difficult to know where to start,* but whatever new genre you settle upon will add to your musical pleasure without dimming the love you have for the music of your youth.

Pattern recognition

One primary aspect of our enjoyment of music is pattern recognition. The human brain is, among other things, an amazingly impressive pattern recognition machine. A pattern involves a certain amount of repetition and predictability, and when most of us think about patterns, the first things that spring to mind are the

* If any of you are in this quandary, you might find the selection of albums from various musical genres at the back of this book useful.

sorts of repeating designs you might get on fabric or wallpaper, like this:

Since the ability to recognize patterns is a very efficient way of making sense of our surroundings, our brain has evolved to make us feel rewarded whenever we successfully identify one. Designs like this one give our brain pleasure because they are predictable, but only if they aren't too boring. We are stimulated by complexities in a pattern or combinations of patterns. Unexpected breaks in, or changes to, patterns also add extra interest and increase arousal.

Nor do patterns have to be perfectly repeating to be pleasing; the branches and thorns on this bush are not precisely evenly spaced, or all the same shape, but our brain categorizes them into a pattern anyway.

Of course, in music we don't experience visual patterns spread out in space like this. Instead we hear patterns that repeat in time.

The most obvious time-based musical pattern is the rhythm. If we hear a pattern such as bom-diddy-bom-bom, bom-diddy-bom-bom, we expect more of the same or perhaps a change to something similar: bom-diddy-diddy-bom, for example. The idea that a rhythm is a pattern is fairly straightforward, but the fact that both the pitch and timbre of any note are also patterns is not so obvious. You might remember this illustration of the pressure waves produced by various instruments from earlier in the book:

flute

oboe

violin

Clearly these are patterns of vibration, which differ from instrument to instrument. The patterns also alter if the pitch of the note changes, and our pattern recognition system monitors all the pitches and timbres of the harmony, as well as the string of notes that make up the melody.

If we combine two or more notes, we change the pattern of vibration of the eardrum, which means that harmonies are also identified by pattern recognition:

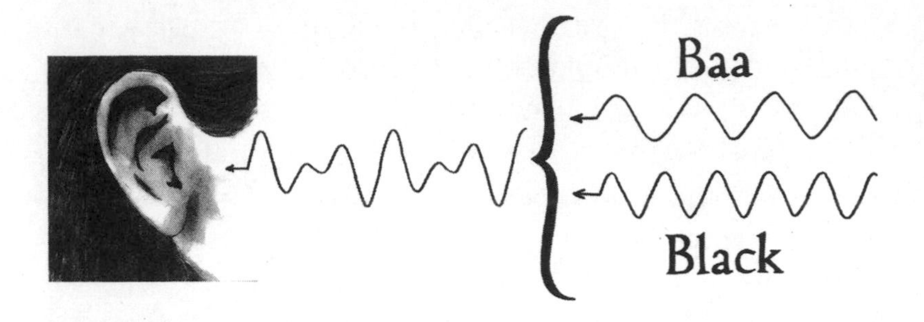

So when we listen to music, our pattern recognition systems are extremely busy identifying the pitch, melody, timbre, harmony, and rhythm of what's being played.

"Very interesting," you might say, "but what has this to do with why I love music?"

All will become clear in a minute, but first I need to explain that we have three types of memory:

- Your *short-term memory* is simply used for storage of what happened over the last few seconds. This is the memory that, as I mentioned in chapter ten, can deal with only about seven different things in play at any one time—which is why musical keys have only about seven notes in them.
- Your *long-term memory* can store any number of things for years on end (but with varying degrees of accuracy).
- Your *working memory* isn't just a storage device; it's an information organizing system that makes sense of the stuff kept in the short-term memory by comparing it with any relevant information filed away in the long-term memory.

Information is taken into the short-term memory and processed in the working memory, where our pattern recognition skills come into play. We humans have a good working memory for things we see and hear. Monkeys, although they have a very good visual working memory, have a poor aural working memory (which is probably why they don't like music). Our highly effective working memory for sounds enables us to deal with long, complicated tunes (or spoken sentences), and is particularly useful when we hear a new piece of music.

When we listen to a piece of music for the first time, our working memory helps us to build up expectations and predictions of what is likely or unlikely to happen next in the rhythm, harmony, timbre, and melody. In the context of music, and from the point of view of a particular individual, a prediction can have one of three outcomes:

1. It can turn out to be correct (e.g., you expected a big, warm final chord to come next, and that's what happened);
2. It can turn out to have been under-optimistic (e.g., you expected a big, warm final chord but the music did something unexpected and delightful); or
3. It can turn out to have been over-optimistic (e.g., you expected a big, warm final chord but something less pleasant happened).

Neuroscientists Robert Zatorre and Valorie Salimpoor have used high-tech brain scanning techniques to analyze what happens inside our heads when we listen to music and found that when we are listening to a new piece of music, a particular area of the brain called the *nucleus accumbens* is very active.[17] This bit of our brain is linked to making predictions, anticipation, and working out whether a particular prediction was eventually correct.

If a prediction turns out to be correct, dopamine is released into our brain (as a reward for predicting the future correctly), and we experience pleasure. Dopamine release is the way we encourage ourselves to repeat behavior that is good for us as individuals and for the human race—like eating and sex. Being good at forecasting things in any context is obviously useful from a survival point of view, so a correct forecast is rewarded by a shot of dopamine.

If a prediction is found to be under-optimistic, then we also get a dopamine release, because things turned even more pleasantly than expected.

If a prediction is over-optimistic, our nucleus accumbens says to itself, "Well, you got that wrong *and* nothing nice happened. No dopamine this time—you don't deserve a reward."

This prediction-based reward-system process can't work for music we are familiar with, because we already know what's going to happen next. However, we *do* experience dopamine rewards when listening to our favorite music, and Drs. Zatorre and Salimpoor have an explanation for this. They found that another area of the brain called the *caudate nucleus* becomes very active when we hear the notes *just before* our favorite bits in music we know well. This part of our brain is associated with anticipation of things we desire. During this anticipation period we get a dose of dopamine, which is actually triggered by the unexciting stuff we hear in the lead-up to our favorite parts of the music. This is similar to the anticipatory pleasure we get as we buy a sandwich when we are hungry. The anticipation mechanism is just as important for survival as the mechanism that assesses how well our predictions are working out. A reward system for forecasting correctly needs to be backed up by an anticipation system that encourages you to keep on doing certain things again and again (eating, drinking, listening to the Rolling Stones, etc.).

Mood-enhancing chemicals

The science fiction writer Iain M. Banks set several of his novels in a highly advanced civilization called "the Culture," the inhabitants of which are rather like upgraded humans. These superior beings have glands that can produce various mood- or performance-enhancing chemicals at will. Just by thinking about it, they can internally generate substances that help them sleep, or stay alert, or calm down. This might strike you as an impossible futuristic idea—but it's not as far-fetched as it sounds. We normal humans have the capacity to secrete similar chemicals; the only difference is that we don't usually have any conscious control over the supply. The chemicals in question are, of course, adrenaline, serotonin, dopamine, and so on. Generally these substances are triggered automatically in response to what's going on around us: if you nearly fall off your bike, you'll get a surge of adrenaline; if your beloved partner kisses you, then you'll get a serotonin hit.

Listening to music is one of the ways in which we can generate these chemicals at will. If you feel the need for some adrenaline to perk you up for a party, try putting on "Baby's on Fire" by Brian Eno. If you need cheering up with a dollop of dopamine and/or serotonin, just play any of your favorite songs and those substances will seep into your brain. Even imagining yourself listening to the piece of music in question can trigger the desired response. All we need now are faster-than-light spaceships, artificial intelligence, and lots of three-legged aliens to battle, and "the Culture" will be ours.

Leaving aside the three-legged aliens, this is a pretty amazing capability. It's taken millennia of evolution and the growth of technology to get to the point where we can actually influence our mental state at the flick of a switch (or play button). This is, of course, why so many of us spend the day plugged in to personal music systems.

And so we come to the end of our journey through the psychology of music.

Since the mid-twentieth century, psychologists and sociologists, musicologists and neurologists, have learned a great deal about the effects music has on us, but there are still a lot of areas that need exploring. In the future we will certainly know more, but I'm sure we'll never run out of questions, and frankly, I think that's a good thing.

Music has the power to alleviate depression, reduce perceived pain, help you cope with various illnesses and disorders, reduce boredom, aid relaxation, help you focus on a physical task, help you bond with others, reduce stress, improve your mood, and fill your life with emotions from nostalgia to joy.

No wonder you love it.

FIDDLY DETAILS

FIDDLY DETAILS A

Timbre

Timbre is so difficult to define in conversational terms that even one of the technical definitions used by professional musicologists is completely upside down:

"**Timbre:** Everything to do with a musical note that is *not* its pitch, loudness, or duration."

The timbre of an instrument is its distinctive voice, and the reason why it's difficult to describe is that it's made up of several components, including:

1. How long the note takes to get going, and the changes in quality of the note during its "start-up" phase. (A flute note starts rapidly; a low trombone note starts more slowly with a sort of "blooming" effect.)
2. What the note sounds like after it's started. (A mid-range flute note sounds clear and pure; the same note on a clarinet sounds richer and more complex.)
3. How the sound of the note changes as it dies away and how quickly it fades. (A harp or xylophone note can take several seconds to fade away unless the player stops the vibration, while a violin or flute note generally dies away in less than a second.)

Every note has a start-up phase, a middle phase, and a dying-away phase, and these are all different, depending on which instrument is playing. So to understand timbre, we need to understand each one of these phases.

First of all we'll look at the most obvious part of the sound, the middle phase: what the note sounds like after it's started.

Let's have another look at those tracings of the repeating pressure patterns produced by different instruments:

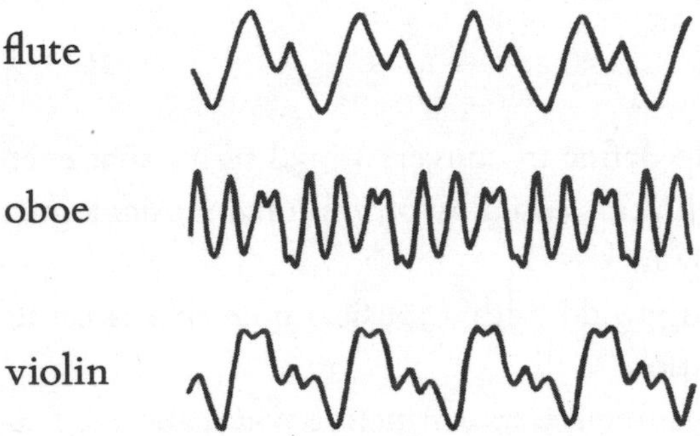

flute

oboe

violin

Just by looking at the patterns here, we can work out three characteristics of these pressure waves:

1. We know that musical notes are made up of repeating patterns of up-and-down pressure, and each pattern here repeats—so they are musical notes.
2. The overall length of a single pressure cycle is the same in each case:

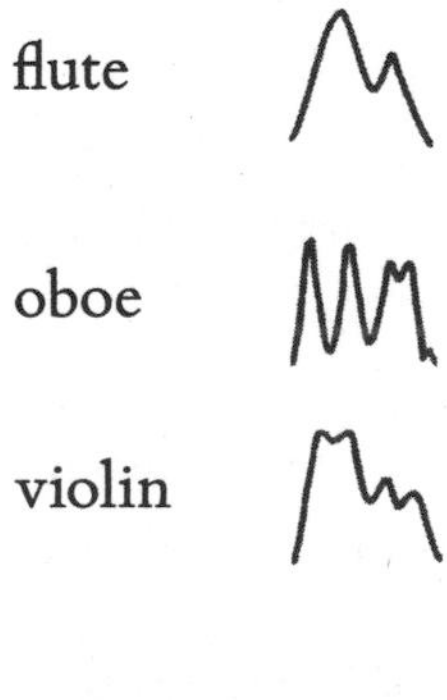

flute

oboe

violin

Since they all have the same cycle time, they all have the same pitch.

3. The pressure waves are all approximately the same height—which means that these notes are of a similar loudness.

Even though we might be listening to the same note played at the same loudness, we can easily discern whether that note is coming from a flute, an oboe, or a violin because the instruments have different *timbres*. As we've seen, the only difference among these three tracings is the detail of the rises and falls of pressure within each complete cycle. These details are the key to the differences in timbre among these instruments.

When specialists in musical acoustics (and other interested parties such as ourselves) want to discuss different timbres, they can't just restrict themselves to saying things like "Well, that clarinet pattern is a hell of a lot lumpier than the flute one." So it's fortunate that there is a useful method of breaking down these complicated repeating patterns into simpler components. To see how it works, let's have a look at how some musical synthesizers imitate real instruments—from clarinets to electric guitars.

The simplest synthesizer notes are made from a pressure wave that goes smoothly up and down in pressure like this:

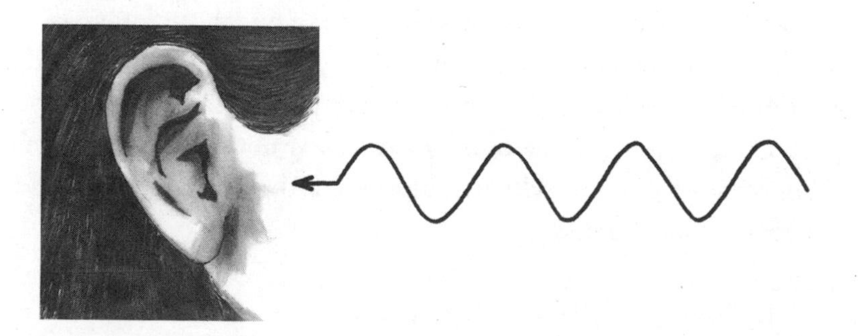

This is called a "pure tone." It is a musical note, but it's a fairly boring one,* and it doesn't sound anything like a real instrument. To make synthesizers sound more like real instruments, you need to add several pure tones together in order to produce more complicated combinations. As you can see in this very simple addition of two waves, things can get complicated quite quickly.

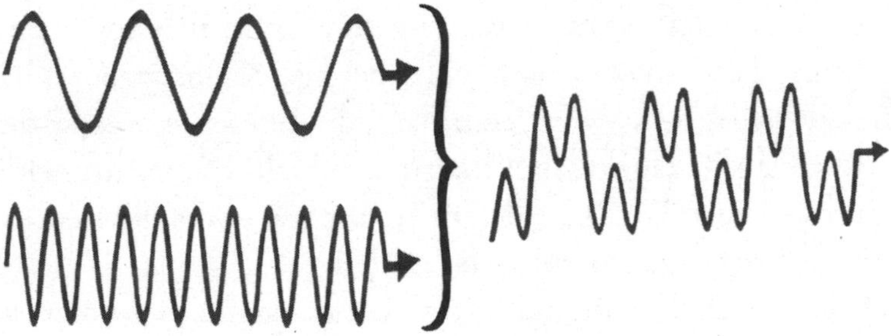

By adding several pure tones together, you can eventually produce an almost exact copy of the wave shapes of real instruments. The flute is one of the simplest wave patterns to copy, which is why early 1970s synthesizers had lots of different flute-like sounds. It took a lot more experimentation before they had a good oboe sound.

But you can't just add any two pure tones together. In this example, I picked a pure tone with a particular cycle time and added a pure tone with one third of that cycle time, meaning that three waves of the second pattern fit exactly into a single wave of the original pure tone. If I'd picked a note in which three-and-a-bit waves of the second note fit into one cycle of the first, then the combined pattern wouldn't repeat exactly each time, and without a repeating pattern we wouldn't have a clear note. We'd have some other kind of noise—which could be anything from an impure-sounding note to a hiss.

* It's also known as a "sine wave."

Here's what it looks like if you add a "three-and-a-bit" pure tone. You can see that the nice repeating pattern has been disrupted:

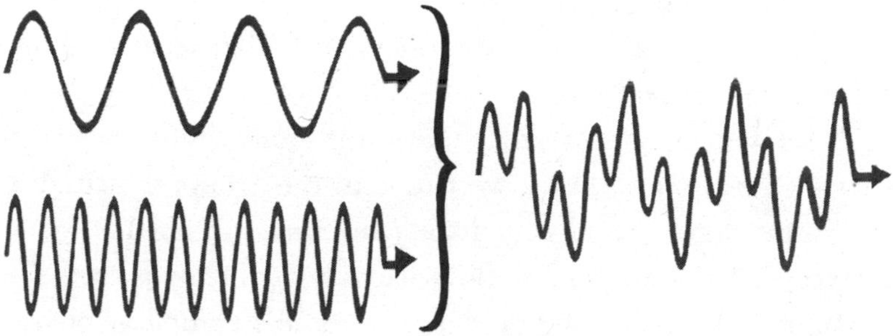

If you want to start adding pure tones together to make notes, you'll need to follow a simple recipe. First, start with your pure tone of choice. Now you can add on:

The pure tone with half this cycle time.
The pure tone with one third of this cycle time.
The pure tone with one quarter of this cycle time.
The pure tone with one fifth of this cycle time.

And so on. But be careful: only fractions with a 1 on top are allowed.

For those of you who prefer to work in frequencies, this is exactly the same as saying that the only ingredients you are allowed are:

A pure tone with a certain frequency
The pure tone with twice that frequency.
The pure tone with three times that frequency.
The pure tone with four times that frequency.
The pure tone with five times that frequency.

And so on.

We'll stay with the word "frequency" from now on because notes are most commonly identified by their frequency—and it's easier to discuss the relationships between multiples (five times, four times, etc.) than fractions.

So, how do combinations of pure tones replicate the timbres of different instruments?

Tones with frequencies that are simple multiples of one another are called *harmonics*. The lowest tone in the group, which all the others are related to, is called the *fundamental* frequency, or first harmonic. The pure tone with twice that frequency is the second harmonic, and so on. The basic idea is that any complex note from a real instrument can be replicated by adding different loudnesses of the various harmonics into the mix.

In the case of the 440 Hz note of the flute, oboe, and violin, the harmonic frequencies are:

440 Hz, 880 Hz, 1320 Hz, 1760 Hz, 2200 Hz, and so on.

We can think of any synthesized note as being produced by a sort of choir in which each choir member sings a single harmonic frequency and the combination of all these frequencies gives us the overall sound of the note.

For the 440 Hz A:

Choir member 1 sings at 440 Hz.
Choir member 2 sings at 880 Hz.
Choir member 3 sings at 1320 Hz.

And so on...

The individual choir members all sing with different loudnesses depending on what instrument is being copied. For example, on a flute, choir member 1 is the loudest, member 3 is the next loudest, and members 2 and 4 (and all the other even numbers) are extremely quiet. On an oboe, the loudest choir member for this note is number 3, followed by 2, 1, and then 4. These different "harmonic choir loudness mixes" give each type of instrument an individual acoustic flavor that is its timbre.

It's also worth bearing in mind that the contribution of the voices to the note changes on any instrument, depending on whether the note is high or low, loud or quiet. For example, the relaxing timbre of a quiet, low saxophone note is quite different from an anxious, loud, high note on the same instrument—because the various choir members have changed their contribution to the note.

Pure tones are very difficult to produce naturally, and we generally need electronics to generate them (although tuning forks produce almost pure tones). Real instruments don't produce a collection of individual pure tone harmonics that are then consolidated into a complicated wave pattern for each note. The fact that real instrument wave shapes can be described as "this much of that harmonic and that much of another harmonic" is just an interesting way of describing what's going on and comparing notes of different timbres.

Let's look at what actually happens in, for example, the case of an acoustic guitar during the course of a single note.

The start-up phase

First of all a string is plucked; it is pulled to one side and then released. It then vibrates backwards and forwards at a certain frequency. Let's say we've picked a string length and tension suitable to give us a frequency of 110 cycles a second (110 Hz). This vibration is passed on to the bridge of the instrument (the thing that attaches the strings to the body of the guitar). The next part of the start-up phase is that the vibration spreads out from the bridge, and the body of the guitar starts to tremble. Any type of vibration that can remain "in step" with the 110 Hz to and fro of the string will be allowed to continue vibrating, but any other type of vibration will quickly die out. (Any vibrations that can't keep in step with 110 Hz tend to get muddled up from one cycle to the next and quickly lose their identity.)

Toward the end of the start-up phase, the sound of the note gets louder, as the vibrations spread to involve more areas of the guitar body, and purer, as all the vibrations that couldn't keep in step with 110 Hz fade away.

The middle phase

Once the note is established, the string, the bridge, the body, and the neck of the instrument are all vibrating in step with the overall frequency of 110 Hz. With the guitar as a whole vibrating like this, there is no chance that the overall vibration could be a simple, pure tone. We therefore get a complicated repeating wave pattern typical of the shape of a wooden acoustic guitar. This is the timbre of a guitar vibrating at this frequency. If you change the shape of the guitar or the material it's made of, then you change the timbre. This means of course that even two supposedly identical guitars have slightly different timbres—because the wood is never identical and there are lots of tiny differences, such as exactly how much glue was used to make them and so on.

The dying-away phase

As the vibrating string loses energy, the vibration changes (just as it did when it gained energy during the start-up phase). Less and less of the guitar is involved in the vibration, and eventually the feebly vibrating string can only pass on energy to the part of the guitar body closest to the bridge. Finally the string stops vibrating and the note dies away completely.

* * *

All notes on acoustic instruments are like this. During the start-up phase the initial vibration is passed on to the rest of the instrument. During the middle phase the instrument carries out a complicated three-dimensional vibration based on the cycle time of the fundamental frequency, and in the final stage the note dies away. During this cycle the shape of the repeating pressure wave being produced undergoes various changes—and the timbre of the instrument is built up not just from the sound of the note in its middle phase but also from the details of the progression from start-up to fade-out. In fact, as far as instrument recognition is concerned, the start-up phase of a note is often as important as the middle phase. As I explained in my book *How Music Works,* if you record an instrument playing a note and then remove the first few milliseconds, you will have an extremely difficult time trying to identify the instrument involved. If you're interested in hearing an example of this effect, just visit my website howmusicworks.info and look for the *How Music Works* CD, track 2: *Can you guess what this instrument is?*

One final point I'd like to make about timbre is that on most instruments the musician has some control over the quality of the sound. For example, a guitarist can pluck the strings close to the bridge, which produces a more twangy, aggressive timbre, and flute players can change the shape of their lips to produce a more breathy sound. Techniques like this alter the "choir mix" of the timbre and are often used to enhance the emotional content of a piece or to make repetitive music more interesting.

FIDDLY DETAILS B

Post-Skip Reversal

Let's take a closer look at rule seven of my rules for melodies from the beginning of chapter ten:

7. After a big step, the next step will probably be smaller and in the reverse direction.

Rule six states that big leaps, or skips, in the pitch of a tune tend to be upwards, and rule seven says that these big upward leaps are usually followed by a smaller step downwards. For example, a big jump up from A to F will probably be followed by a smaller jump downwards, to a note between A and F. Similarly, on the rarer occasions where there is a big leap downwards in pitch, the next note tends to be a smaller step upwards.

This phenomenon is called *post-skip reversal* and it's common throughout much of the world's music. Because it is so widespread, musicians and musicologists have until recently assumed that there is some deep psychological/musical driving force behind post-skip reversal.* Perhaps some emotional longing to return to safer ground after a big leap?

Well, I'm afraid the psychology of music has nothing to do with

* For an example of earlier theories, see L. B. Meyer, *Explaining Music: Essays and Explorations* (University of Chicago Press, 1978).

it. The actual reason why post-skip reversal is so common was discovered by music psychologists Paul von Hippel and David Huron in the late 1990s.[1] After a lot of work studying the folk songs of Europe, South Africa, Native America, and China, they found out that post-skip reversals are simply a matter of probability:

- Each song has a bottom note and a top note.
- The notes used most frequently in a song tend to be somewhere in the middle of this range.
- Just before a large leap upwards, you are likely to be somewhere in the middle of the range.
- So when you hit the top note of your upwards leap, you are likely to be near the top of the range of notes for your tune.
- After the leap you will be producing one of the highest notes in the range of the tune, and there is only a small chance that the next note will involve moving upwards again, to one of the rarely used upper notes. But there is a bigger likelihood of moving downwards to one of the frequently used mid-range notes.
- So a step downwards is very probable.

The same argument is true for a large downward leap: it's likely to be followed by a smaller step upwards.

The statistical principle behind all this is called "regression to the mean," and it holds true in lots of situations. For example, if a shopkeeper has just served a very tall person, the chances are that the next customer will be closer to the average height, simply because very tall people are rare.

In the case of post-skip reversal, the tall person is the equivalent of a high note—and the next note is likely to be closer to the average pitch of the tune.

Von Hippel and Huron then extended their study to look at tunes with big leaps upward that start from a very low note.

According to their "return to mid-range" theory, a big leap upwards from a very low note would not necessarily be followed by a smaller leap downwards. If the leap took you up to a note in the mid-range of the song, then the tune could go either up or down from there. And this, to their great satisfaction, is what they found.

"Baa, Baa, Black Sheep" is an example of this. The first note, "Baa," is the lowest note in the whole song; the next "baa" is the same note again. Then there is a big jump up to "black," but this jump just takes us to the middle of the range of this song, so the next note could be up or down or the same note repeated. In fact the tune repeats the same note for "sheep" and then continues upwards—not a post-skip reversal in sight.

Going back to our shopkeeper, this is the equivalent of serving a very short customer followed by an average-height customer. The third customer is equally likely to be a bit taller, or a bit shorter, or the same height as Mr. Average.

So although post-skip reversal does exist, the cause is not musical/psychological. It all comes down to the statistical fact that notes at the extreme range of any melody are rare (because they are more difficult to sing), and notes in the middle of the range are common.

Only one more of my rules for melodies has a straightforward explanation: rule four states that there are far more small steps than big jumps in most melodies. The obvious reason for this is that small steps are easier to sing than big jumps. Music originated from singing, and so our tunes will tend to involve a lot of easy-to-sing small steps.

FIDDLY DETAILS C

Harmonizing a Tune

As I mentioned in chapter eleven, much of music's emotional impact comes not from the melody itself but from the way it is harmonized with other notes. A change in harmony can completely change the emotional power of a tune.

If you play a carefully chosen sequence of chords along with a tune, not only do you give the music a fuller sound, but you can enhance the melody by supporting the more important notes in it. The harmony can also be used to build up sequences of tension and relaxation.

A full discussion of how we harmonize tunes would take up a whole book (or three), but I think it is important that we discuss a few basic points here. We'll start with how we use chords to support the most important notes in the tune with a standard, relaxed harmony.

If you play "Baa, Baa, Black Sheep" on the piano in the key of C (using only the white notes), starting on middle C (which is also known as C_4), you need to play this sequence of notes to get the tune:

C_4,C_4,G_4,G_4, A_4,B_4,C_5,A_4,G_4, F_4,F_4,E_4,E_4, D_4,D_4,C_4

Matching the notes to the words:

C_4, C_4, G_4, G_4, A_4, B_4, C_5, A_4, G_4, F_4, F_4, E_4, E_4, D_4, D_4, C_4
Baa baa black sheep, have you an - y wool? Yes sir, yes sir, three bags full.

As you can see, the C we use for the "an" in "an-y" is an octave above the one we use for "Baa, baa" and "full."

Fair enough. But what about the chords?

A chord is any combination of three or more different notes played at the same time.* There is no limit to how many notes you can have in your chord. You could, with the help of a few friends, press all the keys on a piano at the same time to produce a (horrible-sounding) eighty-eight-note chord. But most simple chords have only three or four different notes in them.

To get a simple three-note chord from a scale, you choose any note in that scale, then, going up the scale, skip a note and add the next one you come to, then do the same thing again. Like this:

C_2, D_2, E_2, F_2, **$\underline{G_2}$**, A_2, **$\underline{B_2}$**, C_3, **$\underline{D_3}$**, E_3, F_3, G_3, A_3, B_3

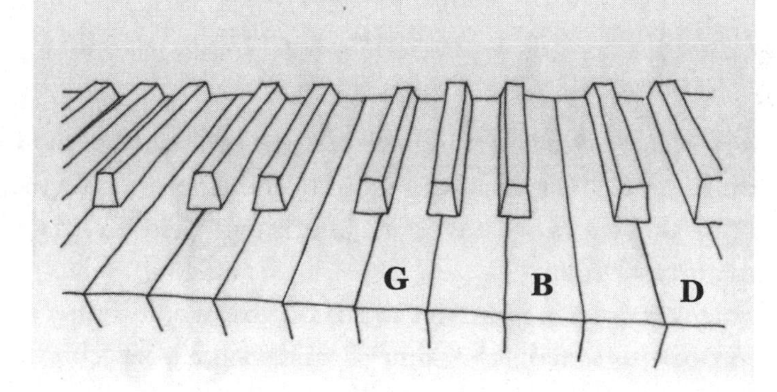

* C_3, C_4, and C_5 played together would not be a chord because, as far as harmony is concerned, notes with the same letter name are similar to one another. For a chord you need three or more different letter names (C, E, and G, for example).

Simple chords like this have names based on the lowest note of the three. In our example we have G, B, and D, which is called G major. For this part of our discussion we'll focus on the three major chords that can be created by any particular major key. In the case of the key of C major these are:

C, E, G—which is the chord of C major
F, A, C—which is the chord of F major

and

G, B, D—which is the chord of G major

These three chords are the cornerstone of most Western music in the key of C major.*

And good old "Baa, Baa, Black Sheep" gives us a typical example of how they can be used.

The first phrase:

Baa	baa	black	sheep
C	C	G	G

would be accompanied by the chord of C major. The tune involves only the notes C and G, and the chord uses the notes C, E, and G, so the tune and the chord have a lot in common. The chord therefore supports the tune.

If you were playing this on a piano, you wouldn't necessarily want to use exactly the same C and G for your chord and your tune. So you might play C_3, E_3, and G_3 with your left hand to make the C major chord, and the C_4 melody note with your right thumb. (A pianist would have all her fingers hovering, ready for the next notes, but I've tucked most of my fingers out of the

* The same principle works for any major key.

way in the photo so you can clearly see which notes are being played.)

Playing a C major chord with your left hand and a C melody note with your right.

Moving on to the second phrase:

have you an - y

A B C A

Accompanying this collection of notes with another C major chord would be a mistake. This is because notes that are next to each other in the scale sound harsh and dissonant if they are played at the same time. In this case the A in the tune would clash with the G in the chord and the B in the tune would clash with the C.

You might have noticed that the C on "**an**-y" *would* be supported by a C major chord—and this brings us to the point about how we choose harmonies that support the important notes in a melody. Melodies often use sequences of small steps between notes (like the A–B–C we have in this example), so whatever chords you use will probably clash with some of the notes involved and support others. Because of this, the general rule is that the chords which sound best are those that favor the notes that are emphasized by the rhythm.

In this case the rhythmically emphasized note is on the word "have." (Try singing the song while exaggerating the emphasis and you'll see what I mean.) Of our three major chords, the best one to suit a melody that goes "A, B, C, A" with the emphasis on the first "A" is F major (F, A, C).

The next note is a long, emphasized G for "wool." This would clash with both the F and the A in the chord of F major, so we need another chord change. We could use either C major (C, E, G) or G major (G, B, D), but C major wins the day as it provides better finality to the end of the phrase. (The keynote chord in any key gives a sense of homecoming and ending—which is why it's often the final chord of a song.)

We can now rewrite the notes for the song with the relevant chords underneath:

Words:	Baa	Baa	Black	Sheep,	Have	You	An-y	Wool?
Tune:	C_4	C_4	G_4	G_4	A_4	B_4	C_5 A_4	G_4
Chords :	G.............................				C........................			G...........
(Each chord	E.............................				A........................			E ....
is made up	C.............................				F........................			C ....
of three notes)								

(The dots after the chords indicate how long they last.)

The same logic is used throughout the rest of the song, and the chord of G major (G, B, D) gets its chance to appear accompanying the repeated D in the melody for the words "three bags."

This is the basic principle of harmonizing in nearly all Western music, from Vivaldi to Madonna. The harmony chords usually include some of the notes of the melody they are accompanying, concentrating on the "important" notes that are rhythmically emphasized. Jazz musicians and modern classical composers sometimes deliberately avoid this cliché—but don't forget: clichés become clichés because they are popular and effective.

Inverted chords

The more eagle-eyed of you readers may have noticed that, although I have put numbers next to the note names for the tune, I have not done so for the chords. This is because of the octave equivalence described in chapter nine. As far as the harmony is concerned, the letters of the notes that make up the chord are far more important than the numbers. Keeping our tune where it is (starting on C_4), you could arrange it so that the chord is lower than the tune, e.g., C_2, E_2, G_2, or higher—C_6, E_6, G_6. You would clearly hear the difference between these two versions, but in both cases the harmony would successfully be doing its job of supporting and emphasizing certain notes and giving more depth to the sound. So the two versions would sound different, but they would both sound pleasant.

Taking this idea to its logical conclusion, there is no need for the notes to be in the order they are—with the C as the lowest note, the E next, and the G at the top. For that first chord C, E, G, we could use it in a standard form—e.g., C_6, E_6, G_6—or we could leave the E and the G where they are and make the C the highest note: E_6, G_6, C_7. If you wanted an extreme case, you could use G_1, C_3, and E_7 and the harmony would still work. This collection of notes is just another version of the chord of C major because it has the C major letters in it.

Chords like this, where the lowest note isn't the standard one for that particular collection of notes, are called *inversions* (or inverted chords) because the chord is sort of upside down.* Inversions are very important when it comes to creating smooth-moving, sophisticated harmonies.

Once again, let's look at our old friend "Baa, Baa" to see how this works:

Using the standard chords we might start with:

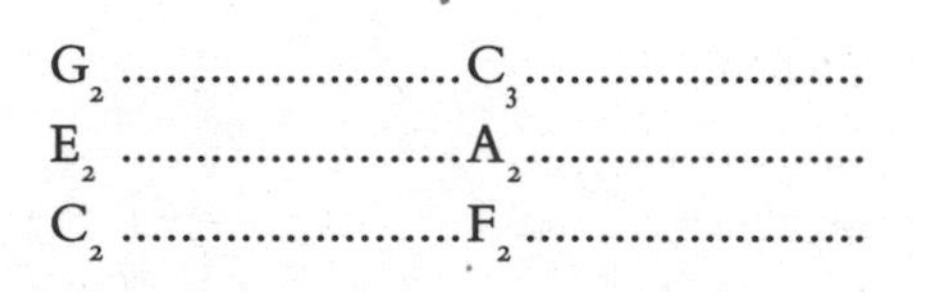

G_2 ....C_3........................
E_2 ....A_2........................
C_2 ....F_2........................

There are circles drawn on the notes that make up a standard C major chord (C, E, G) and exes on the notes of the standard F major chord (F, A, C).

This chord change requires you to put your left hand down on

* Once a chord has been inverted it is no longer named after its lowest note. If you want to find its name, then you need to un-invert it back to its simplest form—as explained at the beginning of this section—and then identify the lowest note.

the piano keyboard for the first chord and then move your hand about three inches to the right for the second chord. No subtlety at all in the movement or the sound, and this harmonization would therefore sound rather abrupt. Now see how the whole thing becomes less clunky if you simply move the C_2 in the first chord up an octave to C_3:

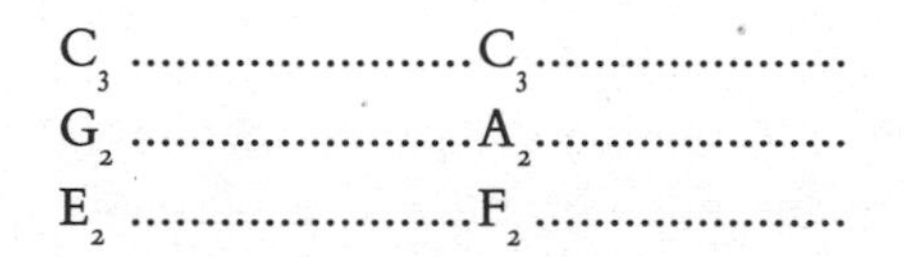

C_3 ....C_3....................
G_2 ....A_2....................
E_2 ....F_2 ....

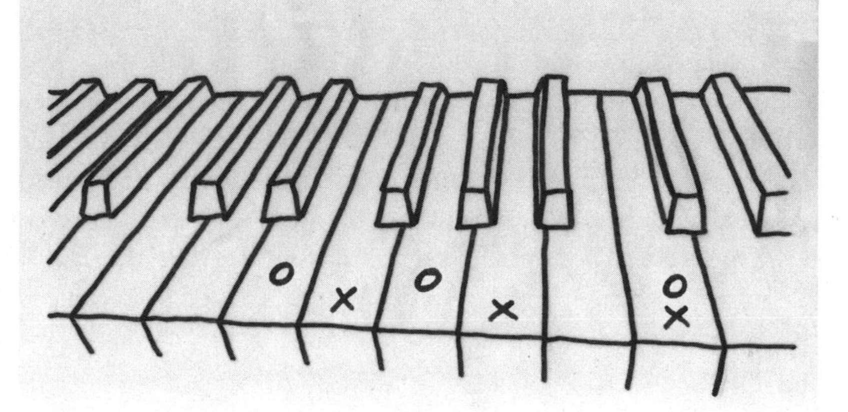

The circles are now on the notes for an alternative (inverted) type of C major chord (E, G, C). The exes are, as in the previous illustration, on the notes of a standard F major chord (F, A, C).

Much smoother. Now the top note stays the same even though the chord changes, and the other two notes move only one note up in each case, the G_2 up to A_2 and the E_2 up to F_2. This is a far more subtle treatment of the harmony, but the chords are basically the same. Hurrah for octave equivalence! (And that's not a sentence you come across very often.)

Techniques like this make it possible to arrange and harmonize even the simplest of tunes in thousands of ways, depending on whether you want subtle gliding from chord to chord, abrupt changes, or a mixture of both.

Making chords more interesting

Playing all the notes of a chord simultaneously is the simplest way to produce a harmony, but you can put a bit more complexity and subtlety into the music by providing the notes of a chord one at a time in a repeating pattern, rather than all together. Chords that have been split up into individual notes like this are called *arpeggios* (meaning harp-like), although guitarists and banjo players often call them "rolls." A lot of bluegrass banjo music consists of a tune interwoven throughout a continuous stream of high-speed arpeggios, and one of the most distinctive uses of arpeggios in pop music is the 1960s hit "The House of the Rising Sun" by the Animals.

Our favorite tune, arranged with arpeggios, could be organized like this:

Baa		baa		black		sheep		have	you	an -	y	wool?			
C		C		G		G		A	B	C	A	G			
		G				G				C				G	
	E		E		E		E		A		A		E		E
C				C				F				C			

Our brain is quite used to this method of receiving a harmony as a sequence of individual notes,* which we subconsciously weave back together into their constituent chords.

---

* I've used very simple arpeggio patterns here, but you can present the notes of the chord in any order, with any rhythm, and the technique still works.

Chords with more than three notes in them

The accompanying chords to a song are often written like this:

C					F
.... On	top	of		Old	Smokey
				C	
All	covered		in	snow	
				G^7	
I	lost	my	true	lover	
				C	
By	courting		too	slow	

Whenever you see songs written down in this way, it's assumed that you know the tune. The letters above certain words mean "start playing this chord at the beginning of this word." In this case we start with a C major chord repeated a few times before the singing starts.* The C major chord is repeated until the singer gets to the beginning of the word "Smokey," when the instrumentalist changes to an F major chord. The F major chord is repeated until the beginning of the word "snow," at which point we switch back to a C major chord... and so on. That little seven next to the G means that an extra note has been added to the G major chord. The additional note is an F, which is the seventh one you come to if you count up from G.

* Capital letters mean that we want a major chord. Minor chords have a lowercase "m" or "min" after the capital letter, like this: Am or Amin.

So the G^7 (G seventh) chord includes these notes:

G, B, D, and F

The addition of the F has a slightly unsettling effect on the G major chord, and this adds to the feeling of relaxation and homecoming we experience when the music changes to the final C major chord.

The song works fine with the chords C major, F major, and G seventh as long as you start on the note C for the words "On top." If you do this, you will later find yourself singing the C an octave above that first note when you sing the "Smo-" in "Smokey." But what happens if that upper C is uncomfortably high for you?

Well, no problem. You can start singing on a lower note. The tune sounds just as good whichever note you start on, as long as you make the correct leaps in pitch as you go along. But if you change your starting note, you'll have to change the chords as well. If you started the tune on G, for example, the correct chords would be G major, C major, and D^7. If you started on E, they would be E major, A major, and B^7.

As I said earlier, harmonization is a huge subject, and the past few pages show only some of the basics of harmonizing a simple melody. Obviously you can create a three-note chord based on each of the seven notes in a major key, and any of these chords could be added occasionally into our three-chord basic mix, but some are far more common than others. A lot of pop/rock songs include a particular chord in a standard progression that (in the key of C major)* goes C major, G major, A minor, and F major. In

* To keep things simple I'm writing here as if everything is written in C major, but this progression works in any key as long as you use the appropriate chords for that key. For example, if the song is in the key of E major, the progression would be E major, B major, C sharp minor, and A major.

their very funny YouTube video "4 chord song" the Australian comedy-rock band Axis of Awesome have demonstrated the pop song ubiquity of this chord progression by repeating it over and over again as the accompaniment to a continuous medley of more than thirty songs (from Bob Marley's "No Woman No Cry" to Lady Gaga's "Poker Face").

Harmony that uses only a small number of standard three-note chords (with the occasional four-note G^7) fills out the sound of the music but doesn't usually add a lot of emotional interest. The tune is the icing on a standard cake mix. Complexity, subtlety, and tension can be introduced by altering one or two of the original three notes, or by adding different notes of a standard chord to the three that form the basis of your chord.

I'm not talking here about adding notes that are simply an octave above or below one of the original three. It's quite common to fill out the sound of a chord by having some or all of the notes appear in your chord several times in different octaves. On a guitar, for example, the chords fill out the music better if you play all six strings—so the standard guitar chord of E major (E, G sharp, B) involves three E's, two B's, and one G sharp. This duplication of notes an octave above or an octave below doesn't count as adding extra notes to the original group as far as the harmony goes. But if, for example, you change one of the E's to an F sharp, then you have added a new note, and created a more complex chord.

Even in straightforward pop songs, four- or five-note chords are pretty common. The extra notes are often considerably less stable team members than the original three, making the chord sound tense and in need of resolution—which will often be provided by the next chord. This tension not only adds musical interest but also tends to drive the music forward as you anticipate the resolution in the same way you look forward to the punch line of a joke.

FIDDLY DETAILS D

How Many Tunes Are Hidden in the Harmony?

I have two boxes in front of me. In the first box there are four pieces of fruit (a banana, an apple, a lemon, and an orange). In the second box there are four useful objects (a pen, a watch, a notebook, and a camera). My job is simply to choose one item from the first box and then choose one item from the second box. When I consider the possibilities, I see that I could have a banana and any one of the four useful objects, or an apple and any of the four, or a lemon and any of the four, or an orange and any of the four. That's four choices multiplied by four choices, so there are in fact a total of sixteen potential combinations.

If I had a third box with four other things in it, the number of possible ways of doing it would be four multiplied by four multiplied by four (a total of sixty-four). As you can see, the number of possibilities rises very rapidly as you add more boxes.

The four flautists I described at the beginning of chapter nine have a job to do. They have to deliver a sequence of nine groups of four notes to the listeners:

Group	1	2	3	4	5	6	7	8	9
	C_5	C_5	G_5	G_5	A_5	B_5	C_6	A_5	G_5
	G_6	G_6	G_6	G_6	F_6	F_6	F_6	F_6	G_6
	E_4	E_4	E_4	E_4	C_5	C_5	C_5	C_5	E_5
	C_4	C_4	C_4	C_4	A_4	A_4	A_4	A_4	C_4

The people in the audience don't care which flautist plays any particular note as long as they hear the correct groups of notes in the right order. Let's imagine that we are going to let the flautists choose their own notes. We'll let flautist 1 have the first pick. She simply has to choose one note from each of the nine groups, and whichever sequence she chooses will be a tune of some sort.

What she is doing is exactly the same process as choosing one of four objects from nine boxes, so the actual number of possible tunes in our simple example is four multiplied by itself nine times: $4 \times 4 \times 4 \times 4 \times 4 \times 4 \times 4 \times 4 \times 4 = 262{,}144$.

* * *

When the sheet music was handed out in the first place, the tune was given to flautist 1, and the boring accompaniment notes were given to the other three musicians. Like this:

	1	2	3	4	5	6	7	8	9
Flute 1	C_5	C_5	G_5	G_5	A_5	B_5	C_6	A_5	G_5
Flute 2	G_6	G_6	G_6	G_6	F_6	F_6	F_6	F_6	G_6
Flute 3	E_4	E_4	E_4	E_4	C_5	C_5	C_5	C_5	E_5
Flute 4	C_4	C_4	C_4	C_4	A_4	A_4	A_4	A_4	C_4

It's important to remember that *any* sequence of notes can be a tune. In fact flautists 2, 3, and 4 are all playing their own tunes. Dull, repetitive tunes—but tunes nonetheless.

These tunes are so dull that our flautists start to rebel. "Why can't we swap a note here and there?" says flautist number 2. "For example, for those first four notes, you've got me playing G_6 four times and my friend flautist 3 is playing E_4 four times. It's really boring for both of us. Why don't we alternate notes? I could play G_6, E_4, G_6, E_4, and flautist 2 could play E_4, G_6, E_4, G_6. It doesn't change what the audience hears but it makes our lives a bit more interesting."

"Yes, good idea!" says flautist 3. "Let's alternate our notes all the way through.

"You play: E_4 G_6 E_4 G_6 C_5 F_6 C_5 F_6 E_5,
and I'll play: G_6 E_4 G_6 E_4 F_6 C_5 F_6 C_5 G_6."

"Hang on," says flautist 4. "That still leaves me bored to death. Why don't we alternate between all three of us? Then it will be even more interesting."

It's not long before the three accompanying flautists realize that their day will get even more exciting if flautist 1 joins in with the note swapping. Eventually the four flautists come to an agreement that they should share all the notes—all the accompaniment notes and the "Baa, Baa, Black Sheep" tune.

After half an hour of negotiation ("In group three I'll swap you my C_4 for your G_6") the flautists have redistributed all the notes and they are all a lot happier because they have chosen not to have any boring repeated notes to play. From the 262,144 possibilities they have eventually chosen this one:

	1	2	3	4	5	6	7	8	9
Flute 1	G_6	E_4	G_5	E_4	A_5	F_6	A_4	C_5	G_5
Flute 2	C_4	G_6	C_4	G_5	C_5	A_4	F_6	A_5	G_6
Flute 3	E_4	C_5	E_4	G_6	F_6	B_5	C_5	F_6	C_4
Flute 4	C_5	C_4	G_6	C_4	A_4	C_5	C_6	A_4	E_5

This looks more complicated than my original version but it is still just four flutes playing the same nine groups of notes in the same order—so an audience wouldn't notice any difference between this and the original. They'd hear the melody line of "Baa, baa, black sheep, have you any wool" and the same simple accompanying harmony.

If I asked you to get a pencil and underline the tune notes in this new distribution scheme, I imagine it would take you two or three minutes of cross-checking with my original version. The amazing thing is that our ears always "underline" the melody notes instantly—with no conscious effort.

Here are the tune notes, underlined and in bold.

	1	2	3	4	5	6	7	8	9
Flute 1	G_6	E_4	$\underline{\mathbf{G}_5}$	E_4	$\underline{\mathbf{A}_5}$	F_6	A_4	C_5	$\underline{\mathbf{G}_5}$
Flute 2	C_4	G_6	C_4	$\underline{\mathbf{G}_5}$	C_5	A_4	F_6	$\underline{\mathbf{A}_5}$	G_6
Flute 3	E_4	$\underline{\mathbf{C}_5}$	E_4	G_6	F_6	$\underline{\mathbf{B}_5}$	C_5	F_6	C_4
Flute 4	$\underline{\mathbf{C}_5}$	C_4	G_6	C_4	A_4	C_5	$\underline{\mathbf{C}_6}$	A_4	E_5

Even with the help of the underlining and bold print it takes some effort to see the tune hidden in among the other notes—but you can't hide a tune from our ears. As I explain in chapter eleven, our hearing system uses the principles of similarity, proximity, good continuation, and common fate to pick out the tune from the accompaniment—in this case spotting that one tune from over a quarter of a million possibilities!

FIDDLY DETAILS E

Scales and Keys

I'll begin this section by taking another look at how we choose the notes for a major scale using the "Just" system. As I explained in chapter twelve, we begin with only two notes: the keynote and the note an octave above it.

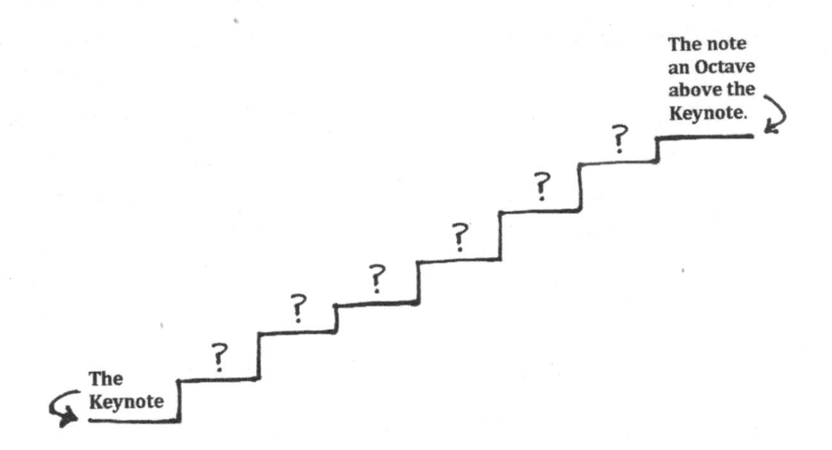

These two notes are in tune with each other because two cycles of the "octave above" note fit exactly into one cycle of the keynote. Another way of saying this is that these two notes are in tune because two halves equal one.

Continuing this line of thinking, it doesn't take long to realize that three thirds are also equal to one. Surely a note that has one third of the keynote's cycle time would also be nicely in tune with the keynote. Wouldn't three of its cycles fit perfectly into the keynote cycle to give us a match that was almost as good as an octave?

This is all very true, but now we have a problem. This new note is even higher in pitch than our "octave above" note. For our scale we need notes that fit inside our octave, and this one is too high.

Happily, octave equivalence comes to the rescue. We know that any two notes an octave apart are linked closely enough to make the note an octave below this "too high" note work pretty well. And so it does. The cycle time of this note is twice one third—that is, two thirds (⅔)—of the keynote cycle time.

Although one cycle of our keynote doesn't match one, two, or three cycles of our new note, we get the next-best thing: two cycles of the keynote match three cycles of the new note, producing a nicely repeating pattern. Because the repeat is spread over two cycles of the keynote, it's not quite as strong a link as the octave coupling I discussed earlier, but we still hear it as a very pleasant combination.

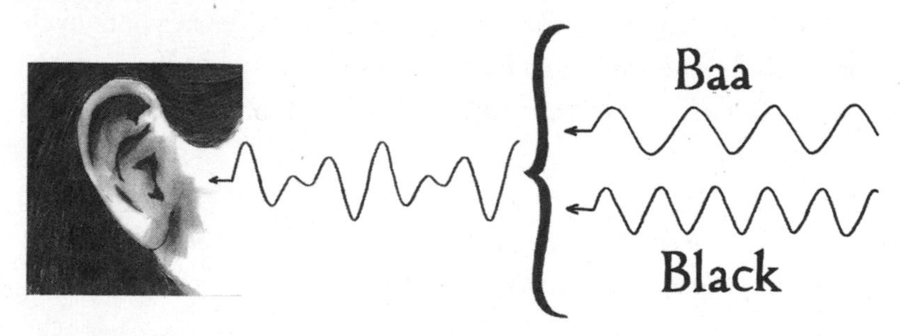

Using the same logic, if we look at the note that has ¾ of the keynote cycle time, we can see that four of its cycles fit exactly into three of the keynote cycles. And sure enough, this does make another nice-sounding pairing.

We now have four notes in our octave and they have the following cycle times:

- the keynote cycle time (the keynote)
- ¾ of the keynote cycle time
- ⅔ of the keynote cycle time
- ½ the keynote cycle time (an octave above the keynote)

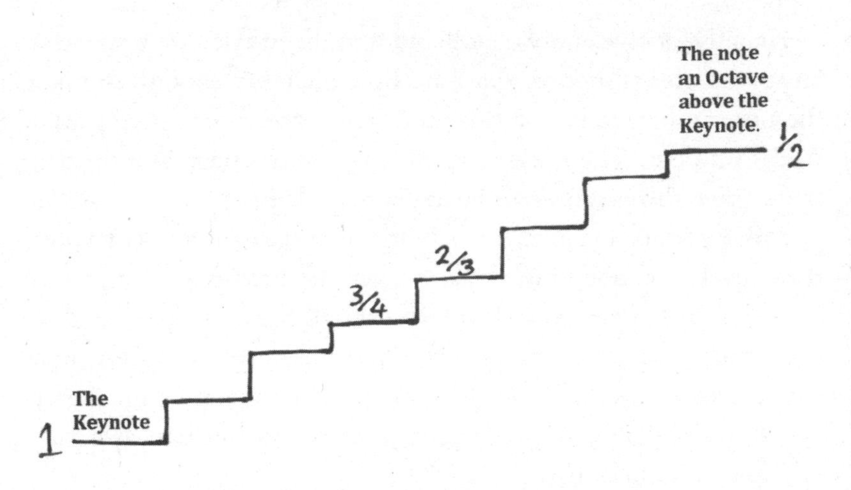

So it's becoming clear that simple fractions of the keynote cycle time might be the answer to finding notes that are in tune with the keynote. In fact we can now come up with a definition of the term "in tune":

> *If a few cycle times of one note fit exactly into a few cycle times of another note, the two notes are in tune with each other.*

And that's how we find all the other notes in our scale. By using simple fractions, we can create a major scale of eight notes that are all as in tune with the keynote as they can be. Here are the fractions of the keynote cycle time for them all, starting with the keynote:

Note 1, the keynote, obviously has the cycle time of the keynote.
Note 2 has a cycle time that is 8/9 of the keynote cycle time.
Note 3 has a cycle time that is 4/5 of the keynote cycle time.
Note 4 has a cycle time that is ¾ of the keynote cycle time.
Note 5 has a cycle time that is ⅔ of the keynote cycle time.
Note 6 has a cycle time that is 3/5 of the keynote cycle time.
Note 7 has a cycle time that is 8/15 of the keynote cycle time.
Note 8 has a cycle time that is ½ the keynote cycle time (the "octave above" note).

If you play any of these notes at the same time as the keynote, they set up a repeating pattern like the ones I drew earlier. The number above the line in each fraction is the number of keynote cycles it takes before the cycle repeats in each case.

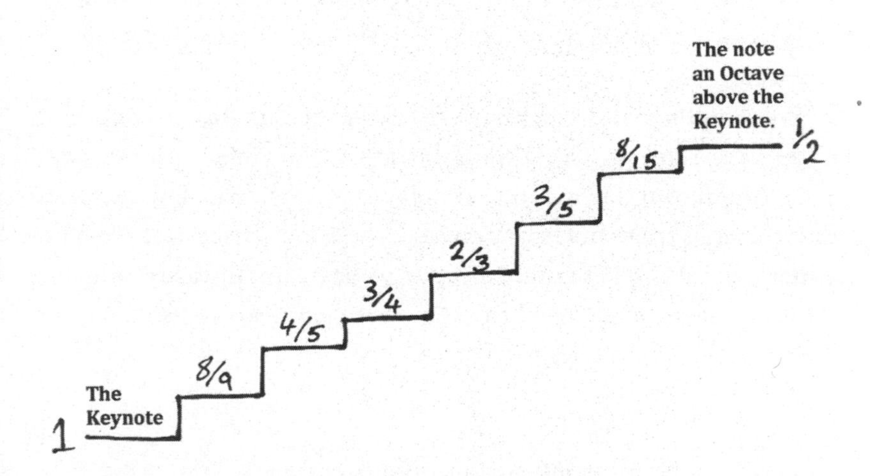

This system of choosing notes from simple fractions creates what is called a Just scale. (It's called "Just" because in the past the word "just" meant "correct" or "true.")

Clearly some of these fractions are simpler than others. When we listen to any of these notes played at the same time as the keynote, we find that the smoothest combinations come from the fractions with the lowest numbers above the fraction line. Our hearing system finds combinations that repeat in less than four or five keynote cycles pleasant, and the lower the number, the smoother the sound. Played at the same time as the keynote, the ½ note sounds smoother than ⅔, which sounds smoother than ¾.

But two of the notes in our scale don't sound smooth at all when they are played at the same time as the keynote. As you might expect, these are the ones with the biggest numbers above the line in their fractions: $^8/_9$ and $^8/_{15}$. In each case we have to wait eight cycles of the keynote before the pattern repeats. Our hearing system struggles to make sense of this, and we hear the combination of the two sounds as rough rather than smooth.

Even though two of the notes in our scale sound rough if they are played at the same time as our keynote, *they are in tune with it,* because a few of their cycles fit exactly into a few cycles of the keynote.

You might wonder why we even bother having a couple of notes in the group that, although they are in tune with the keynote, sound rough when they are played alongside it. The reason is that without these notes there would be a big gap in our scale on either side of the keynote. In order to have small, easily singable steps in our scale, we need to fill in these gaps, so we use the most "in tune" notes we can get to do so.

Problems with the Just scale

On the face of it, the Just scale is an excellent way of dividing up the octave to provide us with a usable number of notes. We have a keynote, the note an octave above, and six notes between them that are in tune with the keynote. Hurrah!

Ah, if only life (and music) were that simple...

Notes 1 (the keynote) and 5 sound great together because the cycle time of note 5 is exactly two thirds of the cycle time of note 1.

Notes 2 and 6 should share the same relationship—but they don't. You might suspect something is different from the fractions involved: the relationship between 1 and ⅔ (notes 1 and 5) is very simple, but the relationship between 8/9 and 3/5 (notes 2 and 6) is obviously more complicated.

Three cycles of note 6 last for a slightly longer time than two cycles of note 2. The difference isn't much, but it's enough to mean that, instead of two cycles of the lower note fitting into three cycles of the upper one, we need *twenty-seven* cycles of the lower note before we get an exact fit.

In this case forty cycles of the higher note fit precisely into twenty-seven cycles of the lower one, but the combined pattern takes so long to repeat that our hearing system assumes that the notes involved don't fit together well enough to pass the "in tune" test. You may remember that to be "in tune," *a few* cycle times of one note need to fit exactly into a few cycle times of another note. Well, twenty-seven is too high a number to be described as "a few," so this combination sounds out of tune.

This is the problem with Just scales and keys: the simple fractions ensure that the notes are all in tune with the keynote, but some of the other notes can be out of tune with one another.*

The Just scale has other intrinsic problems if you use it to tune a

* Those of you who enjoy juggling numbers might like this proof of things not always working out in the Just system. The distance between note 1 and note 3 in a major scale is four semitones, and the distance between note 1 and note 8 is twelve semitones. So three times the four-semitone jump should equal the octave: $4/5 \times 4/5 \times 4/5$ should equal ½. But if you do the calculation, you'll find that the answer (0.512) is slightly more than ½.

piano and you'd like to be able to change key occasionally (which most Western music does). You can't tune a piano (or any other keyboard instrument) to the Just system without some of the chords being out of tune. This is the reason why we use a different tuning system, called equal temperament, for tuning fixed-note instruments.

Equal temperament

Equal temperament (ET) doesn't use fractions to decide on the cycle times or frequencies of the notes in the scale; it uses a mathematical formula to make all the semitone jumps exactly the same size.

The ET system was worked out independently by the Chinese scholar Chu Tsai-Yü in 1580 and by Galileo's father, Vincenzo Galilei, in 1581. The system wasn't commonly used until the late 1700s, though, and the British piano firm Broadwood didn't change over to ET until 1846.

Galilei and Chu Tsai-Yü found that calculating the equal temperament system is pretty easy once you have presented the problem clearly and logically:

1. A note an octave above another must have twice the frequency of the lower one.
2. The octave must be divided up into twelve semitone steps.
3. All the twelve steps must be equal. (If you take any two notes one step apart, then the frequency ratio between them must always be the same.)

To achieve these aims our two wise men calculated that we needed an increase in frequency of 5.9463 percent between two notes which are a semitone apart.* This means that most of the notes in our scale are slightly higher or lower in pitch than they would be in the Just system.

Let's have a look at the difference between a Just and an equal temperament A major chord (which is made up of the three notes A, C sharp, and E). Both the Just and the ET systems use the 440Hz A as their starting point, but the frequencies of the other notes are slightly different for the two systems. For the Just system, the frequency of the E above our A will be exactly 1½ times 440, which is 660 Hz. In the ET system, however, this has been adjusted a bit to 659.2 Hz. The C sharp has been adjusted a larger amount, from 550 Hz up to 554.4.

These differences mean that a Just A major chord sounds very smooth but an ET major chord sounds only moderately smooth. There is some compensation for this, however, because some of the other chords in the Just system sound definitely out of tune, whereas all the chords in the ET system have the same level of moderate smoothness.

Equal temperament has been a source of heated arguments among musicians for over two hundred years. There is an excellent book on the subject by musicologist Ross W. Duffin called *How Equal Temperament Ruined Harmony (and Why You Should Care),* so I think we all know which side of the fence he's on. Yet a lot of the music you've ever heard, particularly solo keyboard music, has been performed using equal temperament tuning, so it's clearly a very successful compromise.

Equal temperament has been by far the most popular tuning

* For example, if A has a frequency of 110 Hz, then the note one semitone up (which we call either A sharp or B flat) has a frequency of 110 multiplied by 1.059463, which is 116.54 Hz. The next note up, B, will have a frequency of 116.54 multiplied by 1.059463, which is 123.47 Hz, and so on.

system for fixed-note Western musical instruments since about 1850, although there are some enthusiasts who still prefer to use various methods of adjusting the Just system to make it work more evenly. If you're interested, have a look at "musical temperament" on YouTube, where you'll find quite a few videos of people comparing various different tuning systems.

But, as I discussed at the end of chapter twelve, although equal temperament is the tuning system we use for fixed-note instruments like keyboards, musical groups that involve only variable-pitch instruments (singers, violinists, etc.) often use Just tuning—particularly for slow, harmonized pieces. Skilled performers on variable-pitch instruments can alter the pitch of the notes they play to compensate for the inherent problems of the Just system and produce purer harmonies than the ones produced by equal temperament.

Acknowledgments

As was the case with *How Music Works,* I would like to take all the credit for the good bits of this book and blame someone else (anyone else) for any omissions, blunders, and stupidities that might be discovered by you readers. Unfortunately this cunning plan is doomed to failure because I have to admit that a lot of people helped me during the course of the past four years.

First of all I would like to thank my excellent, marvelously well-informed editor Tracy Behar and her colleagues Jean Garnett, Genevieve Nierman, and Ben Allen at Little, Brown for all their advice and support. Thanks also to Amanda Heller for her meticulous copyediting.

A host of music specialists and psychologists read various sections of the book and gave me encouragement and feedback. Especial thanks must go to the doyenne of the subject, Professor Diana Deutsch, who very generously spent time reading several chapters, and I would also thank Professors Nicky Dibben, David Hargreaves, John Sloboda, and Sandra Trehub.

My heroic girlfriend, Kim, gets a gold medal for reading and correcting about ten versions of the manuscript and giving calming and expert technical assistance whenever my computer did anything particularly stupid or malicious.

And my greatest thanks must go to my friend Dr. John Wykes, who provided a constant stream of good advice, great information, and stern instructions to pull my socks up and try to explain things properly!

Without all the excellent help and advice I received from these people (and my friends Rod O'Connor, Pierre Lafrance, Jim Carpenter, Whit, John Dingle, Patricia Lancaster, Mini Grey, Jo Grey, Bob and Jen Malone, Clem Young, David Osborne, and DJ Fresh), I wouldn't have been able to put together the good bits of the book.

And as for those omissions, blunders, and stupidities, I want to make it quite clear that they have nothing to do with me. I think it's only reasonable that, as usual, I attribute any blame for such things to my former secondary school geography teacher, Mr. Nigel Jones, of 14B Mountebank Close, Eccles, Lancashire. Please send any negative comments, lawsuits, or demands for compensation directly to him at the above address.

Cheers,
John
Direct email:
howmusicworks@yahoo.co.uk

Finally, I would like to also thank the following two firms for giving me permission to quote song lyrics:

Lyrics from "Bad Moon Rising" written by John Cameron Fogerty, reproduced by permission of Concord Music Group, Inc.

Lyrics to Bob Dylan's "Make You Feel My Love" reproduced by kind permission of Bob Dylan Music Co.

Suggestions for Listening and Viewing

A. Listening

Here is a very short list of stuff you might enjoy listening to. I've restricted myself to five each of classical, jazz, and world music, because otherwise the list would go on forever—and anyway I don't think anyone would be interested in my selection of progressive rock from the 1970s. I hope you enjoy them.

World Music

1. Various artists, *Real World 25*. A selection of the past twenty-five years of world music recordings from Real World Records. This is my top recommendation of all.
2. Madredeus, *O Paraíso*. A lyrical acoustic Portuguese band with an amazing female singer.
3. Ry Cooder and V. M. Bhatt, *A Meeting by the River*. East meets West on slide guitars.
4. Various artists, *The Very Best of Éthiopiques*. African jazz—a great collection. (There are a couple of albums of this name. The one I have in mind is on the Manteca label, with a photo of someone pointing upward on the cover.)
5. Nusrat Fateh Ali Khan, *Mustt*. Brilliant Pakistani singer.

Jazz

1. Miles Davis, *Kind of Blue*. Slinky jazz on trumpet, sax, piano, bass, and drums.
2. Keith Jarrett, *The Köln Concert*. Improvised jazz piano.
3. Jim Hall and Bill Evans, *Undercurrent*. Piano and guitar duets of jazz standards.
4. E.S.T., *Seven Days of Falling*. Twenty-first-century jazz trio.
5. Django Reinhardt and Stéphane Grappelli, *Souvenirs*. Hot swing from Paris.

Classical

1. Joaquín Rodrigo, *Concierto de Aranjuez*. This is the most famous guitar concerto of them all. My favorite performance of it is the one guitarist John Williams made in 1965 with members of the Philadelphia Orchestra conducted by Eugene Ormandy. On CD you can often get it combined with Rodrigo's *Fantasía para un Gentilhombre,* which is also excellent. Cheerful, sunny music from Spain.
2. W. A. Mozart, Clarinet Concerto. Often combined on CD with his Oboe Concerto and Bassoon Concerto.
3. R. Vaughan Williams, *Fantasia on a Theme of Thomas Tallis*. Often combined on CD with other relaxing stuff like Vaughan Williams's *The Lark Ascending*. I like the CD played by the Academy of St. Martin in the Fields, conducted by Neville Marriner (ARGO 414595-2).
4. J. S. Bach, complete lute suites. My favorite performance is the CD by American guitarist Sharon Isbin (Virgin Classics 0777 7595032 8).
5. John Adams, *Harmonium*. John Adams combines modern minimalist techniques with tunefulness. My favorite recording is the one with the San Francisco Symphony orchestra conducted by Edo De Waart (ECM 821 465-2).

B. Viewing

Here is a very short collection of stuff you might find interesting, amusing, or impressive that I've stumbled across (or been told about) on YouTube.

1. The world's oldest musical instrument (40,000 BCE).
   Mammoth ivory and bird bone flutes from Germany, as discussed at the beginning of chapter fifteen.
2. The Concert for George, Anoushka Shankar, HD, parts 2–4.
   Anoushka plays sitar and conducts an orchestra of Western and Eastern instruments and singers. Eric Clapton plays guitar.
3. KT Tunstall "Black Horse and the Cherry Tree," Jools Holland debut, RAVE, HD.
   KT Tunstall's TV debut—just her voice, a guitar, a tambourine, and a virtuoso display of looping. (The looper is the electric gizmo on the floor that she treads on to record what she's playing now to add it to what she plays next, to build up layer after layer of sound.)
4. Just intonation versus equal temperament.
   Different scale systems.
5. Extraordinaire instrument de musique.
   Fun animation.

References and Further Reading

A. Main textbook references and suggestions for further reading

Chapter 1 What Is Your Taste in Music?

Adrian North and David Hargreaves, "Musical Preference and Taste," chap. 3 of *The Social and Applied Psychology of Music* (Oxford University Press, 2008).

Peter J. Rentfrow and Jennifer A. McDonald, "Preference, Personality, and Emotion," chap. 24 of *Handbook of Music and Emotion: Theory, Research, Applications,* ed. Patrick Juslin and Justin Sloboda (Oxford University Press, 2010).

Chapter 2 Lyrics, and Meaning in Music

Adrian North and David Hargreaves, *The Social and Applied Psychology of Music* (Oxford University Press, 2008).

Chapter 3 Music and Your Emotions

Patrik Juslin and John Sloboda, "Music and Emotion," chap. 15 of *The Psychology of Music,* 3rd ed., ed. Diana Deutsch (Academic Press, 2013).

John Sloboda, "Music in Everyday Life: The Role of Emotions," chap. 18 of *Handbook of Music and Emotion: Theory, Research, Applications,* ed. Patrik Juslin and John Sloboda (Oxford University Press, 2010).

Chapter 4 Repetition, Surprises, and Goose Bumps

Elizabeth Helmuth Margulis, *On Repeat: How Music Plays the Mind* (Oxford University Press, 2014).

David Huron, *Sweet Anticipation: Music and the Psychology of Expectation* (MIT Press, 2006).

Patrik Juslin and John Sloboda, "Music and Emotion," chap. 15 of *The Psychology of Music,* 3rd ed., ed. Diana Deutsch (Academic Press, 2013).

Chapter 5 Music as Medicine

Suzanne B. Hanser, "Music, Health and Well-Being," chap. 30 of *Handbook of Music and Emotion: Theory, Research, Applications,* ed. Patrik Juslin and John Sloboda (Oxford University Press, 2010).

Adrian North and David Hargreaves, "Music, Business, and Health," chap. 5 of *The Social and Applied Psychology of Music* (Oxford University Press, 2008).

Chapter 6 Does Music Make You More Intelligent?

E. Glenn Schellenberg and Michael W. Weiss, "Music and Cognitive Abilities," chap. 12 of *The Psychology of Music,* 3rd ed., ed. Diana Deutsch (Academic Press, 2013).

Chapter 7 From Psycho *to* Star Wars: *The Power of Movie Music*

Kathryn Kalinak, *Film Music: A Very Short Introduction* (Oxford University Press, 2010).

Siu-Lan Tan, Annabel Cohen, Scott Lipscombe, and Roger Kendall, eds., *The Psychology of Music in Multimedia* (Oxford University Press, 2013).

Annabel Cohen, "Music as a Source of Emotion in Film," chap. 31 of *Handbook of Music and Emotion: Theory, Research, Applications,* ed. Patrik Juslin and John Sloboda (Oxford University Press, 2010).

Chapter 8 Are You Musically Talented?

Geoff Colvin, *Talent Is Overrated* (Nicholas Brealey Publishing, 2008).

Adrian North and David Hargreaves, "Composition and Musicianship," chap. 2 of *The Social and Applied Psychology of Music* (Oxford University Press, 2008).

Isabelle Peretz, "The Biological Foundations of Music: Insights from Congenital Amusia," chap. 13 of *The Psychology of Music,* 3rd ed., ed. Diana Deutsch (Academic Press, 2013).

Chapter 9 A Few Notes about Notes

John Powell, *How Music Works: The Science and Psychology of Beautiful Sounds, from Beethoven to the Beatles and Beyond* (Little, Brown and Company, 2010).

Charles Taylor, *Exploring Music: The Science and Technology of Tones and Tunes* (IOP Publishing, 1992).

Ian Johnston, *Measured Tones* (IOP Publishing, 1989).

Chapter 10 What's in a Tune?

David Huron, *Sweet Anticipation: Music and the Psychology of Expectation* (MIT Press, 2006).
Aniruddh D. Patel and Steven M. Demorest, "Comparative Music Cognition: Cross-Species and Cross-Cultural Studies," chap. 16 of *The Psychology of Music*, 3rd ed., ed. Diana Deutsch (Academic Press, 2013).
David Temperley, *Music and Probability* (MIT Press, 2010).

Chapter 11 Untangling the Tune from the Accompaniment

Diana Deutsch, "Grouping Mechanisms in Music," chap. 6 of *The Psychology of Music*, 3rd ed., ed. Diana Deutsch (Academic Press, 2013).

Chapter 12 Don't Believe Everything You Hear

William Forde Thompson, "Intervals and Scales," chap. 4 of *The Psychology of Music*, 3rd ed., ed. Diana Deutsch (Academic Press, 2013).
John A. Sloboda, "Music, Language, and Meaning," chap. 2 of *The Musical Mind: The Cognitive Psychology of Music* (Oxford University Press, 1985).

Chapter 13 Dissonance

Isabelle Peretz, "Towards a Neurobiology of Musical Emotions," chap. 5 of *Handbook of Music and Emotion: Theory, Research, Applications*, ed. Patrik Juslin and John Sloboda (Oxford University Press, 2010).
William Forde Thompson, "Intervals and Scales," chap. 4 of *The Psychology of Music*, 3rd ed., ed. Diana Deutsch (Academic Press, 2013).

Chapter 14 How Musicians Push Our Emotional Buttons

Andreas C. Lehman, John A. Sloboda, and Robert H. Woody, *Psychology for Musicians* (Oxford University Press, 2007).
Eric Clarke, Nicola Dibben, and Stephanie Pitts, *Music and Mind in Everyday Life* (Oxford University Press, 2010).
Bruno Nettl, "Music of the Middle East," chap. 3 of Bruno Nettl et al., *Excursions in World Music*, 2nd ed. (Prentice-Hall, 1997).
Reginald Massey and Jamila Massey, "Ragas," chap. 10 of *The Music of India* (Stanmore Press, 1976).

Chapter 15 Why You Love Music

Marjorie Shostak, *Nisa: The Life and Words of a !Kung Woman* (Routledge, 1990).

Laurel J. Trainor and Erin E. Hannon, "Musical Development," chap. 11 of *The Psychology of Music,* 3rd ed., ed. Diana Deutsch (Academic Press, 2013).

Raymond MacDonald, David J. Hargreaves, and Dorothy Miell, "Musical Identities," chap. 43 of *The Oxford Handbook of Music Psychology,* ed. Susan Hallam, Ian Cross, and Michael Thaut (Oxford University Press, 2009).

B. References

In the following list of references you'll find details of the original work published by psychologists, sociologists, musicologists, and many other -ologists of various persuasions, which I used as source material for this book. If any of the researchers involved find that I've quoted their work incorrectly, or without acknowledging them, please accept my profuse apologies and contact me ASAP so I can correct the error in future editions.

Chapter 1 What Is Your Taste in Music?

1. Peter J. Rentfrow and Jennifer A. McDonald, "Preference, Personality and Emotion," chap. 24 of *Handbook of Music and Emotion: Theory, Research, Applications,* ed. P. N. Juslin and J. A. Sloboda (Oxford University Press, 2010), sec. 24.2.
2. See reference 1, section 24.3.2.
3. P. J. Rentfrow and S. D. Gosling, "The Do Re Mi's of Everyday Life: The Structure and Personality Correlates of Music Preferences," *Journal of Personality and Social Psychology* 84 (2003): 1236–56.
4. Adrian North and David Hargreaves, "Musical Preference and Taste," chap. 3 of *The Social and Applied Psychology of Music* (Oxford University Press, 2008).
5. M. B. Holbrook and R. M. Schindler, "Age, Sex and Attitude towards the Past as Predictors of Consumers' Aesthetic Tastes for Cultural Products," *Journal of Marketing Research* 31 (1994): 412–422. See also M. B. Holbrook, "An Empirical Approach to Representing Patterns of Consumer Tastes, Nostalgia, and Hierarchy in the Market for Cultural Products," *Empirical Studies of the Arts* 13 (1995): 55–71. And M. B. Holbrook and R. M. Schindler, "Commentary on 'Is There a Peak in Popular Music Preference at a Certain Song-Specific Age? A Replication of Holbrook and Schindler's 1989 Study,'" *Musicae Scientiae* 17, no. 3 (2013): 305–308.
6. A. C. North, D. J. Hargreaves, and S. A. O'Neill, "The Importance of Music to Adolescents," *British Journal of Educational Psychology* 70, no. 2 (June 2000): 255–272.
7. Carl Wilson, *Let's Talk about Love: A Journey to the End of Taste* (Bloomsbury, 2007), p. 91.
8. North and Hargreaves, "Musical Preference and Taste" (see reference 4).
9. Vladimir Konecni and Dianne Sargent-Pollock, "Choice between Melodies

Differing in Complexity under Divided-Attention Conditions," *Journal of Experimental Psychology: Human Perception and Performance* 2 (1976): 347–356.

10. R. G. Heyduk, "Rated Preference for Musical Composition as It Relates to Complexity and Exposure Frequency," *Perception and Psychophysics* 17 (1975): 84–91; See also North and Hargreaves, "Musical Preference and Taste" (see reference 4).
11. A. N. North and D. Hargreaves, "Responses to Music in Aerobic Exercise and Yogic Relaxation Classes," *British Journal of Psychology* 87, no. 4 (1996): 535–547.
12. Adrian North, David Hargreaves, and Jennifer McKendrick, "The Influence of In-Store Music on Wine Selections," *Journal of Applied Psychology* 84, no. 2 (1999): 271–276. See also Adrian North and David Hargreaves, "Music, Business, and Health," chap. 5 of *The Social and Applied Psychology of Music* (Oxford University Press 2008).
13. Charles Areni and David Kim, "The Influence of Background Music on Shopping Behaviour: Classical versus Top Forty Music in a Wine Store," *Advances in Consumer Research* 20 (1993): 336–340.
14. Adrian North, "Wine and Song: The Effect of Background Music on the Taste of Wine," *British Journal of Psychology* 103, no. 3 (August 2012): 293–301.
15. Peter Stone and Susan Hickey, "Sound Matters," documentary on Irish radio station RTE, broadcast December 10, 2011.
16. Ronald Milliman, "Using Background Music to Affect the Behavior of Supermarket Shoppers," *Journal of Marketing* 46 (1982): 86–91.
17. Ronald Milliman, "The Influence of Background Music on the Behavior of Restaurant Patrons," *Journal of Consumer Research* 133 (1986): 286–289.
18. Clare Caldwell and Sally Hibbert, "The Influence of Music Tempo and Musical Preference on Restaurant Patrons' Behavior," *Psychology and Marketing* 19 (2002): 895–917.
19. T. C. Robally, C. McGreevy, R. R. Rongo, M. L. Schwantes, P. J. Steger, M. A. Wininger, and E. B. Gardner, "The Effect of Music on Eating Behavior," *Bulletin of the Psychonomic Society* 23 (1985): 221–222.
20. North and Hargreaves, "Music, Business, and Health" (see reference 12). See also Adrian North and David Hargreaves, "The Effect of Music on Atmosphere and Purchase Intentions in a Cafeteria," *Journal of Applied Social Psychology* 28 (1998): 2254–73.
21. North and Hargreaves, "Music, Business, and Health" (see reference 12), p. 290.

Chapter 2 Lyrics, and Meaning in Music

1. V. Stratton and A. H. Zalanowski, "Affective Impact of Music vs. Lyrics," *Empirical Studies of the Arts* 12 (1994): 129–140.
2. Vladimir Konecni, "Elusive Effects of Artists' 'Messages,'" in *Cognitive Processes in the Perception of Art,* ed. W. R. Crosier and A. J. Chapman (Elsevier, 1984), pp. 71–93.

3. J. Leming, "Rock Music and the Socialisation of Moral Values in Early Adolescence," *Youth and Society* 18 (June 1987): 363–383.
4. William Drabkin, notes to mini-score, Beethoven Symphony no. 6, Eulenburg ed., no. 407 (Eulenburg 2011), p. 8.
5. Eric F. Clarke, "Jimi Hendrix's 'Star Spangled Banner,'" chap. 2 of *Ways of Listening: An Ecological Approach to the Perception of Musical Meaning* (Oxford University Press, 2005).
6. Jostein Gaarder, *Sophie's World: A Novel about the History of Philosophy* (1991; Orion, 2000).
7. Adrian North and David Hargreaves, *The Social and Applied Psychology of Music* (Oxford University Press, 2008), p. 79.
8. Philip Ball, *The Music Instinct: How Music Works and Why We Can't Do Without It* (Bodley Head, 2010), p. 241.

Chapter 3 Music and Your Emotions

1. N. H. Frijda, "The Laws of Emotion," *American Psychologist* 43, no. 5 (May 1988): 349–358.
2. Patrik Juslin and John Sloboda, "Music and Emotion," chap. 15 of *The Psychology of Music,* 3rd ed., ed. Diana Deutsch (Academic Press, 2013).
3. Robert Plutchik, *Emotions and Life: Perspectives from Psychology, Biology, and Evolution* (American Psychological Association, 2003).
4. Deryck Cooke, *The Language of Music* (Oxford University Press, 1959), p. 115.
5. See reference 4, p. 55.
6. W. F. Thomson and B. Robitaille, "Can Composers Express Emotions through Music?" *Empirical Studies of the Arts* 10 (1992): 79–89.
7. P. N. Juslin, S. Liljestrom, P. Laukka, D. Vastfjall, and L.-O. Lundqvist, "Emotional Reactions to Music in a Nationally Representative Sample of Swedish Adults: Prevalence and Causal Influences," *Musicae Scientiae* 15 (July 2011): 174–207.
8. L.-O. Lundqvist, F. Carlsson, P. Hilmersson, and P. N. Juslin, "Emotional Responses to Music: Experience, Expression and Physiology," *Psychology of Music* 37 (2009): 61–90.
9. Juslin and Sloboda, "Music and Emotion" (see reference 2).
10. Lars Kuchinke, Herman Kappelhoff, and Stefan Koelsch, "Emotion in Narrative Films: A Neuroscientific Perspective," chap. 6 of *The Psychology of Music in Multimedia,* ed. Siu-Lan Tan, Annabel Cohen, Scott Lipscombe, and Roger Kendall (Oxford University Press, 2013).
11. Anne Blood and Robert Zatorre, "Intensely Pleasurable Responses to Music Correlate with Activity in Brain Regions Implicated in Reward and Emotion," *Proceedings of the National Academy of Sciences* 98 (September 2001): 11818–23.

12. Stefan Koelsch, T. Fritz, D. Y. von Cramon, K. Muller, and A. D. Friederici, "Investigating Emotion with Music: An FMRI Study," *Human Brain Mapping* 27 (2006): 239–250.
13. Juslin and Sloboda, "Music and Emotion" (see reference 2).
14. Juslin and Sloboda, "Music and Emotion" (see reference 2).
15. P. N. Juslin, S. Liljestrom, D. Vastfjall, G. Barradas, and A. Silva, "An Experience Sampling Study of Emotional Reactions to Music: Listener, Music and Situation," *Emotion* 8 (2008): 668–683.
16. William Forde Thompson, *Music, Thought, and Feeling: Understanding the Psychology of Music,* 2nd ed. (Oxford University Press, 2015), p. 179.
17. John Sloboda, *Exploring the Musical Mind: Cognition, Emotion, Ability, Function* (Oxford University Press, 2004), pp. 319–331.
18. John Sloboda, "Music in Everyday Life: The Role of Emotions," chap. 18 of *Handbook of Music and Emotion: Theory, Research, Applications,* ed. Patrik Juslin and John Sloboda (Oxford University Press, 2010).
19. E. Bigand, S. Filipic, and P. Lalitte, "The Time Course of Emotional Response to Music," *Annals of the New York Academy of Science* 1060 (2005): 429–437.
20. Sloboda, "Music in Everyday Life: The Role of Emotions" (see reference 18), p. 495.
21. J. H. McDermott and M. D. Hauser, "Nonhuman Primates Prefer Slow Tempos but Dislike Music Overall," *Cognition* 104 (2007): 654–668.
22. Aniruddh Patel and Steven Demorest, "Comparative Music Cognition: Cross-Species and Cross-Cultural Studies," chap. 16 of Deutsch, *The Psychology of Music* (see reference 2).
23. K. E. Gfeller, "Musical Components and Styles Preferred by Young Adults for Aerobic Fitness Activities," *Journal of Music Therapy* 25 (1988): 28–43. See also Suzanne B. Hanser: "Music, Health and Well-Being," chap. 30 of Juslin and Sloboda, *Handbook of Music and Emotion* (see reference 18).
24. Sloboda, "Music in Everyday Life" (see reference 18), p. 508. See also John Sloboda, A. M. Lamont, and A. E. Greasley, "Choosing to Hear Music: Motivation, Process and Effect," in *The Oxford Handbook of Music Psychology,* ed. Susan Hallam, Ian Cross, and Michael Thaut (Oxford University Press, 2009), pp. 431–440.
25. Elizabeth Margulis, "Attention, Temporality, and Music That Repeats Itself," chap. 3 of *On Repeat: How Music Plays the Mind* (Oxford University Press, 2014).
26. T. Shenfield, S. E. Trehub, and T. Nakota, "Maternal Singing Modulates Infant Arousal," *Psychology of Music* 31 (2003): 365–375.
27. J. M. Standley, "The Effect of Music-Reinforced Non-nutritive Sucking on Feeding Rate of Premature Infants," *Journal of Paediatric Nursing* 18, no. 3 (June 2003): 169–173.
28. Juslin and Sloboda, "Music and Emotion" (see reference 2).

29. John A. Sloboda, *The Musical Mind: The Cognitive Psychology of Music* (Oxford University Press, 1985), p. 213.
30. S. Dalla Bella, I. Peretz, L. Rousseau, and N. Gosselin, "A Developmental Study of the Affective Value of Tempo and Mode in Music," *Cognition* 80 (2001): B1–B10.
31. D. Huron and M. J. Davis, "The Harmonic Minor Scale Provides an Optimum Way of Reducing Average Melodic Interval Size, Consistent with Sad Affect Cues," *Empirical Musicology Review* 7, no. 3–4 (2012): 103–117.
32. Patrik Juslin and Petri Laukka, "Communication of Emotions in Vocal Expression and Music Performance: Different Channels, Same Code?" *Psychological Bulletin* 129, no. 5 (2003): 770–814. See also Aniruddh Patel, *Music, Language, and the Brain* (Oxford University Press, 2010). And E. Coutinho and N. Dibben, "Psychoacoustic Cues to Emotion in Speech Prosody and Music," *Cognition and Emotion* (2012), DOI:10.1080/02699931.2012.732559.
33. Dwight Bolinger, *Intonation and Its Parts: Melody in Spoken English* (Stanford University Press, 1986).
34. Thompson, *Music, Thought, and Feeling* (see reference 16), p. 315.
35. David Huron, "The Melodic Arch in Western Folksongs," *Computing in Musicology* 10 (1996): 3–23.
36. Diana Deutsch, "The Processing of Pitch Combinations," chap. 7 of Deutsch, *The Psychology of Music* (see reference 2).
37. Patel, *Music, Language, and the Brain* (see reference 32).
38. G. Ilie and W. F. Thompson, "A Comparison of Acoustic Cues in Music and Speech for Three Dimensions of Affect," *Music Perception* 23 (2006): 310–329.
39. Eric Clarke, Nicola Dibben, and Stephanie Pitts, *Music and Mind in Everyday Life* (Oxford University Press, 2010), pp. 18–19.
40. Deutsch, "The Processing of Pitch Combinations" (see reference 36).
41. Juslin and Sloboda, "Music and Emotion" (see reference 2). See also Patrik N. Juslin, Simon Liljestrom, Daniel Västfjäll, and Lars-Olov Lundqvist, "How Does Music Evoke Emotions? Exploring the Underlying Mechanisms," chap. 22 of Juslin and Sloboda, *Handbook of Music and Emotion*.

Chapter 4 Repetition, Surprises, and Goose Bumps

1. Elizabeth Helmuth Margulis, *On Repeat: How Music Plays the Mind* (Oxford University Press, 2014), pp. 15–16.
2. Peter Kivy, *The Fine Art of Repetition: Essays in the Philosophy of Music* (Cambridge University Press, 1993), p. 356.
3. Margulis, *On Repeat* (see reference 1), p. 15.
4. J. S. Horst, K. L. Parsons, and N. M. Bryan, "Get the Story Straight: Contextual Repetition Promotes Word Learning from Storybooks," *Frontiers in Psychology* 2, no. 17, DOI:10.3389/fpsyg.2011.00017.

5. D. Deutsch, T. Henthorn, and R. Lapidis, "Illusory Transformation from Speech to Song," *Journal of the Acoustical Society of America* 129, no. 4 (April 2011): 2245–52.
6. R. Brochard, D. Abecasis, D. Potter, R. Ragot, and C. Drake, "The 'Tick-tock' of Our Internal Clock: Direct Brain Evidence of Subjective Accents in Isochronous Sequences," *Psychological Science* 14, no. 4 (July 2003): 362–366. See also David Huron, "Expectation in Time," chap. 10 of *Sweet Anticipation: Music and the Psychology of Expectation* (MIT Press, 2006).
7. Robert B. Zajonc, "Attitudinal Effects of Mere Exposure," *Journal of Personality and Social Psychology,* monograph suppl. 9 no. 2, pt. 2 (June 1968): 1–27.
8. David Huron, "Surprise," chap. 2 of *Sweet Anticipation: Music and the Psychology of Expectation* (MIT Press, 2006).
9. Huron, "Surprise" (see reference 8), p. 38.
10. C. He, L. Hotson, and L. J. Trainor, "Development of Infant Mismatch Responses to Auditory Pattern Changes between 2 and 4 Months Old," *European Journal of Neuroscience* 29, no. 4 (February 2009): 861–867. See also Laurel J. Trainor and Robert Zatorre, "The Neurobiological Basis of Musical Expectations," chap. 16 of *The Oxford Handbook of Music Psychology,* ed. Susan Hallam, Ian Cross, and Michael Thaut (Oxford University Press, 2009), pp. 431–440.
11. Huron, "Surprise" (see reference 8).
12. John Sloboda, "Music Structure and Emotional Response: Some Empirical Findings," *Psychology of Music* 19 (1991): 110–120. See also Patrik Juslin and John Sloboda, "Music and Emotion," chap. 15 of *The Psychology of Music,* 3rd ed., ed. Diana Deutsch (Academic Press, 2013).
13. R. R. McCrae, "Aesthetic Chills as a Universal Marker of Openness to Experience," *Motivation and Emotion* 31, no. 1 (2007): 5–11.
14. Sloboda, "Music Structure and Emotional Response" (see reference 12).
15. L. R. Bartel, "The Development of the Cognitive-Affective Response Test—Music." *Psychomusicology* 11 (1992): 15–26.

Chapter 5 Music as Medicine

1. For serotonin, see S. Evers and B. Suhr, "Changes in Neurotransmitter Serotonin but Not of Hormones during Short Time Music Perception," *European Archives of Psychiatry and Clinical Neuroscience,* no. 250 (2000): 144–147; for dopamine, see V. Menon and D. Levitin, "The Rewards of Music Listening: Response and Physiological Connectivity in the Mesolimbic System," *Neuroimage* 28 (2005): 175–184.
2. M. H. Thaut and B. L. Wheeler, "Music Therapy," chap. 29 of *Handbook of Music and Emotion: Theory, Research, Applications,* ed. Patrik Juslin and John Sloboda (Oxford University Press, 2010).
3. S. B. Hanser and L. W. Thompson, "Effects of a Music Therapy Strategy on Depressed Older Adults," *Journal of Gerontology* (1994): 49 265–269. See

also S. B. Hanser, "Music, Health and Well-Being," chap. 30 of *Handbook of Music and Emotion: Theory, Research, Applications,* ed. Patrik Juslin and John Sloboda (Oxford University Press, 2010).

4. Hanser, "Music, Health and Well-Being" (see reference 3), p. 868.
5. C. J. Brown, A. Chen, and S. F. Dworkin, "Music in the Control of Human Pain," *Music Therapy* 8 (1989): 47–60.
6. Adrian North and David Hargreaves, "Music, Business, and Health," chap. 5 of *The Social and Applied Psychology of Music* (Oxford University Press, 2008), pp. 301–311. See also L. A. Mitchell, R. A. R. MacDonald, C. Knussen, and M. A. Serpell, "A Survey Investigation of the Effects of Music Listening on Chronic Pain," *Psychology of Music* 35 (2007): 39–59. And William Forde Thompson, *Music, Thought, and Feeling: Understanding the Psychology of Music,* 2nd ed. (Oxford University Press, 2015), p. 220.
7. L. A. Mitchell and R. A. R. MacDonald, "An Experimental Investigation of the Effects of Preferred and Relaxing Music on Pain Perception," *Journal of Music Therapy* 63 (2006): 295–316.
8. Oliver Sacks, "Speech and Song: Aphasia and Music Therapy," chap. 16 in *Musicophilia: Tales of Music and the Brain* (Knopf, 2007).
9. Hanser, "Music, Health and Well-Being" (see reference 3), p. 870.
10. Sacks, *Musicophilia* (see reference 8), p. 252.
11. North and Hargreaves, "Music, Business, and Health" (see reference 6), p. 308.
12. S. A. Rana, N. Akhtar, and A. C. North, "Relationship between Interest in Music, Health and Happiness," *Journal of Behavioural Sciences* 21, no. 1 (June 2011): 48.
13. Laszlo Harmat, Johanna Takacs, Robert Bodizs, "Music Improves Sleep Quality in Students," *Journal of Advanced Nursing* 62, no. 3 (2008): 327–335.
14. Hui-Ling Lai and Marion Good, "Music Improves Sleep Quality in Older Adults," *Journal of Advanced Nursing* 49, no. 3 (2005): 234–244.
15. D. Chadwick and K. Wacks, "Music Advance Directives: Music Choices for Later Life," 11th World Congress of Music Therapy, Brisbane, Australia, July 2005, cited in Hanser, "Music, Health and Well-Being" (see reference 3).

Chapter 6 Does Music Make You More Intelligent?

1. F. H. Rauscher, G. L. Shaw, and K. N. Ky, "Music and Spatial Task Performance," *Nature* 365 (October 1993): 611.
2. Adrian North and David Hargreaves discuss "the Mozart effect" in "Composition and Musicianship," chap. 2 of *The Social and Applied Psychology of Music* (Oxford University Press, 2008), pp. 70–74.
3. J. Pietschnig, M. Voracek, and A. K. Formann, "Mozart Effect–Shmozart Effect: A Meta-Analysis," *Intelligence* 38 (2008): 314–323.
4. K. M. Nantais and E. G. Schellenberg, "The Mozart Effect: An Artifact of Preference," *Psychological Science* 10 (1999): 370–373. See also E. Glenn Schellenberg and Michael W. Weiss, "Music and Cognitive Abilities," chap. 12

of *The Psychology of Music,* 3rd ed., ed. Diana Deutsch (Academic Press, 2013).

5. M. Isen, "A Role for Neuropsychology in Understanding the Facilitating Influence of Positive Affect on Social Behavior and Cognitive Processes," in *The Oxford Handbook of Positive Psychology,* 2nd ed., ed. S. J. Lopez and C. R. Snyder (Oxford University Press, 2011), pp. 503–518.
6. F. G. Ashby, A. M. Isen, and A. U. Turken, "A Neuropsychological Theory of Positive Affect and Its Influence on Cognition," *Psychological Review* 106 (1999): 355–386.
7. W. F. Thompson, E. G. Schellenberg, and G. Hussain, "Arousal, Mood, and the Mozart Effect," *Psychological Science* 12 (2001): 248–251.
8. E. G. Schellenberg and S. Hallam, "Music Listening and Cognitive Abilities in 10 and 11 Year Olds: The Blur Effect," *Annals of the New York Academy of Sciences* 1060 (2005): 202–209.
9. William Forde Thompson, *Music, Thought, and Feeling: Understanding the Psychology of Music,* 2nd ed. (Oxford University Press, 2015), p. 302. See also Gabriela Ilie and William Forde Thompson, "Experiential and Cognitive Changes Following Seven Minutes Exposure to Music and Speech," *Music Perception* 28, no. 3 (February 2011): 247–264.
10. D. Miscovic, R. Rosenthal, U. Zingg, D. Metzger, and L. Janke, "Randomized Control Trial Investigating the Effect of Music on the Virtual Reality Laparoscopic Learning Performance of Novice Surgeons," *Surgical Endoscopy* 22 (208): 2416–20.
11. K. Kallinen, "Reading News from a Pocket Computer in a Distracting Environment: Effects of the Tempo of Background Music," *Computers in Human Behavior* 18 (2002): 537–551.
12. Daniel Kahneman, *Thinking Fast and Slow* (Farrar, Straus and Giroux, 2013), p. 55.
13. C. F. Lima and S. L. Castro, "Speaking to the Trained Ear: Musical Expertise Enhances the Recognition of Emotions in Speech Prosody," *Emotion* 11 (2011): 1021–31.
14. S. Moreno, E. Bialystok, R. Barac, E. G. Schellenberg, N. J. Cepeda, and T. Chau, "Short-Term Music Training Enhances Verbal Intelligence and Executive Function," *Psychological Science* 22 (2011): 1425–33.
15. L. M. Patston and L. J. Tippett, "The Effect of Background Music on Cognitive Performance in Musicians and Non-musicians," *Music Perception* 29 (2011): 173–183.
16. D. Southgate and V. Roscigno, "The Impact of Music on Childhood and Adolescent Achievement," *Social Science Quarterly* 90 (2009): 13–21.
17. J. Haimson, D. Swain, and E. Winner, "Are Mathematicians More Musical Than the Rest of Us?" *Music Perception* 29 (2011): 203–213.
18. E. G. Schellenberg, "Music Lessons Enhance IQ," *Psychological Science* 15 (2004): 511–514.

Chapter 7 From **Psycho** *to* **Star Wars:** *The Power of Movie Music*

1. Annabel Cohen, "Music as a Source of Emotion in Film," chap. 31 of *Handbook of Music and Emotion: Theory, Research, Applications,* ed. Patrik Juslin and John Sloboda (Oxford University Press, 2010).
2. Kathryn Kalinak, *Film Music: A Very Short Introduction* (Oxford University Press, 2010), p. 44.
3. Cohen, "Music as a Source of Emotion in Film" (see reference 1).
4. W. F. Thompson, F. A. Russo, and D. Sinclair, "Effects of Underscoring on the Perception of Closure in Filmed Events," *Psychomusicology* 13 (1994): 9–27.
5. M. G. Boltz, "Musical Soundtracks as a Schematic Influence on the Cognitive Processing of Filmed Events," *Music Perception* 18, no. 4 (2001): 427–454, cited in David Bashwiner, "Musical Analysis for Multimedia: A Perspective from Music Theory," chap. 5 of *The Psychology of Music in Multimedia,* ed. Siu-Lan Tan, Annabel Cohen, Scott Lipscombe, and Roger Kendall (Oxford University Press, 2013).
6. Cohen, "Music as a Source of Emotion in Film" (see reference 1); A. J. Cohen, K. A. MacMillan, and R. Drew, "The Role of Music, Sound Effects and Speech on Absorption in a Film: The Congruence-Associationist Model of Media Cognition," *Canadian Acoustics* 34 (2006): 40–41.
7. Kalinak, *Film Music* (see reference 2), p. 27.
8. Kalinak, *Film Music* (see reference 2), pp. 14–15.
9. R. Y. Granot and Z. Eitan, "Musical Tension and the Interaction of Dynamic Auditory Parameters," *Music Perception* 28 (2011): 219–245.
10. Zohar Eitan, "How Pitch and Loudness Shape Musical Space and Motion," chap. 8 of Tan, Cohen, Lipscombe, and Kendall, *The Psychology of Music in Multimedia* (see reference 5).
11. William Forde Thompson, *Music, Thought, and Feeling: Understanding the Psychology of Music,* 2nd ed. (Oxford University Press, 2015), p. 162.
12. M. G. Boltz, M. Schulkind, and S. Kantra, "Effects of Background Music on the Remembering of Filmed Events," *Memory and Cognition* 19, no. 6 (1991): 593–606, cited in Berthold Hoeckner and Howard Nusbaum, "Music and Memory in Film and Other Multimedia: The Casablanca Effect," chap. 11 of Tan, Cohen, Lipscombe, and Kendall, *The Psychology of Music in Multimedia* (see reference 5).
13. E. S. Tan, *Emotion and the Structure of Narrative Film: Film as an Emotion Machine* (Routledge, 1995).
14. E. Eldar, O. Ganor, R. Admon, A. Bleich, and T. Hendler, "Feeling the Real World: Limbic Response to Music Depends on Related Content," *Cerebral Cortex* 17 (2007): 2828–40, cited in Annabel Cohen, "Congruence Association Model of Music and Multimedia: Origin and Evolution," chap. 2 of Tan, Cohen, Lipscombe, and Kendall, *The Psychology of Music in Multimedia.*

15. Kalinak, *Film Music* (see reference 2), p. 63.
16. Roger Kendall and Scott Lipscombe, "Experimental Semiotics Applied to Visual, Sound and Musical Structures," chap. 3 of Tan, Cohen, Lipscombe, and Kendall, *The Psychology of Music in Multimedia*.
17. "Diegetic Music, Non-diegetic Music and Source Scoring," www.filmmusicnotes.com, posted April 21, 2013, by Film Score Junkie.
18. Kalinak, *Film Music* (see reference 2), p. 102.
19. Hoeckner and Nusbaum, "Music and Memory in Film and Other Multimedia" (see reference 12).
20. Kalinak, *Film Music* (see reference 2), p. 105.
21. Sandra K. Marshall and Annabel J. Cohen, "Effects of Musical Soundtracks on Attitudes toward Animated Geometric Figures," *Music Perception* 6, no. 1 (Fall 1988): 95–112.
22. Cohen, "Congruence Association Model" (see reference 14).
23. Carolyn Bufford, "The Psychology of Film Music," Psychologyinaction.org, posted November 5, 2012.
24. Kendall and Lipscombe, "Experimental Semiotics Applied to Visual, Sound and Musical Structures" (see reference 16).
25. L. A. Cook and D. L. Valkenburg, "Audio-Visual Organization and the Temporal Ventriloquism Effect between Grouped Sequences: Evidence that Unimodal Grouping Precedes Cross-Modal Integration," *Perception* 38, no. 8 (2009): 1220–33.
26. G. L. Fain, *Sensory Transduction* (Sinauer Associates, 2003).
27. Scott Lipscombe, "Cross-Modal Alignment of Accent Structures in Multimedia," chap. 9 of Tan, Cohen, Lipscombe, and Kendall, *The Psychology of Music in Multimedia* (see reference 5), p. 196.
28. Herbert Zettl, *Sight, Sound, Motion: Applied Media Aesthetics,* 7th ed. (Wadsworth Publishing Co., 2013), p. 80.
29. Marilyn Boltz, "Music Videos and Visual Influences on Music Perception and Appreciation: Should You Want Your MTV?" chap. 10 of Tan, Cohen, Lipscombe, and Kendall, *The Psychology of Music in Multimedia* (see reference 5).
30. Lipscombe, "Cross-Modal Alignment of Accent Structures in Multimedia" (see reference 27), p. 208.

Chapter 8 Are You Musically Talented?

1. John A. Sloboda, Jane W. Davidson, Michael J. A. Howe, and Derek G. Moore, "The Role of Practice in the Development of Performing Musicians," *British Journal of Psychology* 87 (1996): 287–309.
2. K. Anders Ericsson, Ralf Th. Krampe, and Clemens Tesch-Romer, "The Role of Deliberate Practice in the Acquisition of Expert Performance,"

Psychological Review 100, no. 3 (1993): 363–406, cited in Geoff Colvin, *Talent Is Overrated* (Nicholas Brealey Publishing, 2008), pp. 57–61.

3. Anthony Kemp and Janet Mills, "Musical Potential," in *The Science and Psychology of Musical Performance,* ed. Richard Parncut and Gary E. McPherson (Oxford University Press, 2002), pp. 3–16.
4. Sloboda, Davidson, Howe, and Moore, "The Role of Practice in the Development of Performing Musicians" (see reference 1), p. 301.
5. J. W. Davidson, M. J. A. Howe, D. G. Moore, and J. A. Sloboda, "The Role of Teachers in the Development of Musical Ability," *Journal of Research in Music Education* 46 (1998): 141–160.
6. S. Hallam and V. Prince, "Conceptions of Musical Ability," *Research Studies in Music Education* 20 (2003): 2–22.
7. Isabelle Peretz, "The Biological Foundations of Music: Insights from Congenital Amusia," chap. 13 of *The Psychology of Music,* 3rd ed., ed. Diana Deutsch (Academic Press, 2013).

Chapter 10 What's in a Tune?

1. Paul Roberts, *Images: The Piano Music of Claude Debussy* (Amadeus Press, 1996), p. 121.
2. David Huron, "Statistical Properties of Music," chap. 5 of *Sweet Anticipation: Music and the Psychology of Expectation* (MIT Press, 2006).
3. David Temperley, *Music and Probability* (MIT Press, 2010), p. 58.
4. Huron, "Statistical Properties of Music" (see reference 2), p. 65.
5. Huron, "Statistical Properties of Music" (see reference 2), p. 195.
6. Temperley, *Music and Probability* (see reference 3), p. 147.
7. David Temperley, "Revision, Ambiguity, and Expectation" chap. 8 of *The Cognition of Basic Musical Structures* (MIT Press, 2004).
8. Diana Deutsch, "The Processing of Pitch Combinations," chap. 7 of *The Psychology of Music,* 3rd ed., ed. Diana Deutsch (Academic Press, 2013).
9. C. Liegeois, I. Peretz, M. Babei, V. Laguitton, and P. Chauvel, "Contribution of Different Cortical Areas in the Temporal Lobes to Music Processing," *Brain* 121 (1998): 1853–67.
10. Aniruddh Patel and Steven Demorest, "Comparative Music Cognition: Cross-Species and Cross-Cultural Studies," chap. 16 of Deutsch, *The Psychology of Music* (see reference 8).
11. P. C. M. Wong, A. K. Roy, and E. H. Margulis, "Bimusicalism: The Implicit Dual Enculturation of Cognitive and Affective Systems," *Music Perception* 27 (2009): 291–307.
12. S. M. Demorest, S. J. Morrison, L. A. Stambaugh, M. N. Beken, T. L. Richards, and C. Johnson, "An fMRI Investigation of the Cultural Specificity of Musical Memory," *Social Cognitive and Affective Neuroscience* 5 (2010): 282–291.

13. H. H. Stuckenschmidt, *Schoenberg: His Life, World and Work,* trans. Humphrey Searle (Schirmer Books, 1977), p. 277.
14. Howard Goodall, *The Story of Music* (Chatto and Windus, 2013), p. 219.
15. G. A. Miller. "The Magical Number Seven, Plus or Minus Two: Some Limits on Our Capacity for Processing Information," *Psychological Review* 63, no. 2 (March 1956): 81–97.
16. Anthony Pople, *Berg: Violin Concerto* (Cambridge University Press, 1991), pp. 39, 40, 65, and passim.
17. Keith Richards and James Fox, *Life* (Weidenfeld and Nicolson, 2011), p. 511.
18. R. Brauneis, "Copyright and the World's Most Popular Song," GWU Legal Studies Research Paper no. 392, *Journal of the Copyright Society of the USA* 56, no. 355 (2009).
19. See reference 18, p. 11.
20. T. S. Eliot, "Philip Massinger," in *The Sacred Wood: Essays on Poetry and Criticism,* Bartleby.com.

Chapter 11 Untangling the Tune from the Accompaniment

1. Diana Deutsch, "Grouping Mechanisms in Music," chap. 6 of *The Psychology of Music,* 3rd ed., ed. Diana Deutsch (Academic Press, 2013).

Chapter 12 Don't Believe Everything You Hear

1. John Backus, *The Acoustical Foundations of Music,* 2nd ed. (W. W. Norton and Co., 1977), pp. 238–239.
2. J. Chen, M. H. Woollacott, S. Pologe, and G. P. Moore, "Pitch and Space Maps of Skilled Cellists: Accuracy, Variability, and Error Correction," *Experimental Brain Research* 188, no. 4 (July 2008): 493–503. See also J. Chen, M. H. Woollacott, and S. Pologe, "Accuracy and Underlying Mechanisms of Shifting Movements in Cellists," *Experimental Brain Research* 174, no. 3 (2006): 467–476.
3. John A. Sloboda, *The Musical Mind: The Cognitive Psychology of Music* (Oxford University Press, 1985), p. 23.
4. A. M. Liberman, K. S. Harris, J. A. Kinney, and H. Lane, "The Discrimination of the Relative Onset Time of the Components of Certain Speech and Non-speech Patterns," *Journal of Experimental Psychology* 61 (1961): 379–388.
5. S. Locke and L. Kellar, "Categorical Perception in a Non-linguistic Mode," *Cortex* 9, no. 4 (December 1973): 355–369.
6. J. A. Siegel and W. Siegel, "Categorical Perception of Tonal Intervals: Musicians Can't Tell Sharp from Flat," *Perception and Psychophysics* 21, no. 5 (1977): 399–407. See also William Forde Thompson, "Intervals and Scales," chap. 4 of *The Psychology of Music,* 3rd ed., ed. Diana Deutsch (Academic Press, 2013).
7. B. C. J. Moore, "Frequency Difference Limens for Short-Duration Tones," *Journal of the Acoustical Society of America* 54 (1973): 610–619.

8. C. Micheyl, K. Delhommeau, X. Perrot, and A. J. Oxenham, "Influence of Musical and Psychoacoustical Training on Pitch Discrimination," *Hearing Research* 219 (2006): 36–47.

Chapter 13 Dissonance

1. A. J. Blood, R. J. Zatorre, P. Bermudez, and A. C. Evans, "Emotional Responses to Pleasant and Unpleasant Music Correlate with Activity in Paralimbic Brain Regions," *Nature Neuroscience* 2 (1999): 382–387.
2. L. J. Trainor and B. M. Heinmiller,"The Development of Evaluative Responses to Music: Infants Prefer to Listen to Consonance over Dissonance," *Infant Behavior and Development* 21 (1998): 77–88. See also William Forde Thompson, "Intervals and Scales," chap. 4 of *The Psychology of Music,* 3rd ed., ed. Diana Deutsch (Academic Press, 2013).
3. C. Chiandetti and G. Vallortigara, "Chicks Like Consonant Music," *Psychological Science* 22 (2011): 1270–73.
4. Isabelle Peretz, "Towards a Neurobiology of Musical Emotions," chap. 5 of *Handbook of Music and Emotion: Theory, Research, Applications,* ed. Patrik Juslin and John Sloboda (Oxford University Press, 2010).

Chapter 14 How Musicians Push Our Emotional Buttons

1. Andreas C. Lehman, John A. Sloboda, and Robert H. Woody, *Psychology for Musicians* (Oxford University Press, 2007), p. 85.
2. J. A. Sloboda, "Individual Differences in Music Performance," *Trends in Cognitive Sciences* 4, no. 10 (October 2000): 397–403.
3. A. Penel and C. Drake, "Sources of Timing Variations in Music Performance: A Psychological Segmentation Model," *Psychological Research* 61 (1998): 12–32.
4. E. Istok, M. Tervaniemi, A. Friberg, and U. Seifert, "Effects of Timing Cues in Music Performances on Auditory Grouping and Pleasantness Judgments," conference paper, Tenth International Conference on Music Perception and Cognition, Sapporo, Japan, August 25–29, 2008.
5. Lehman, Sloboda, and Woody, *Psychology for Musicians* (see reference 1), p. 97.
6. Eric Clarke, Nicola Dibben, and Stephanie Pitts, *Music and Mind in Everyday Life* (Oxford University Press, 2010), p. 40.
7. B. H. Repp, "The Aesthetic Quality of a Quantitatively Average Music Performance: Two Preliminary Experiments," *Music Perception* 14 (1997): 419–444.
8. Richard Ashley, "All His Yesterdays: Expressive Vocal Techniques in Paul McCartney's Recordings," unpublished manuscript, referenced in Lehman, Sloboda, and Woody, *Psychology for Musicians,* p. 90.
9. Howard Goodall, *The Story of Music* (Chatto and Windus, 2013), p. 304.

10. David Huron, "Creating Tension," chap. 15 of *Sweet Anticipation: Music and the Psychology of Expectation* (MIT Press, 2006), p. 324.
11. Ashley Kahn, *Kind of Blue: The Making of the Miles Davis Masterpiece* (Granta Books, 2001), p. 29.
12. *Last Word* (obituary program), BBC Radio 4, December 7, 2012.
13. Bruno Nettl, "Music of the Middle East," chap. 3 of Bruno Nettl et al., *Excursions in World Music,* 2nd ed. (Prentice-Hall, 1997), p. 62.
14. Reginald Massey and Jamila Massey, "Ragas," chap. 10 of *The Music of India* (Stanmore Press, 1976), p. 104.
15. Kathryn Kalinak, *Film Music: A Very Short Introduction* (Oxford University Press, 2010), p. 13.
16. David Huron, "Mental Representation of Expectation (II)," chap. 12 of *Sweet Anticipation* (see reference 10), p. 235.
17. K. L. Scherer and J. S. Oshinsky, "Cue Utilization in Emotion Attribution from Auditory Stimuli," *Motivation and Emotion* 1, no. 4 (1977): 331–346.
18. Lehmann, Sloboda, and Woody, *Psychology for Musicians* (see reference 1), p. 86.

Chapter 15 Why You Love Music

1. N. J. Conrad, M. Malina, and S. C. Munzel, "New Flutes Document the Earliest Musical Tradition in South-Western Germany," *Nature* 460, no. 7256 (2009): 737–740.
2. Jared Diamond, "Farmer Power," chap. 4 of *Guns, Germs and Steel: A Short History of Everybody for the Last 13,000 Years* (Vintage, 2000), p. 86.
3. Peter Gray, "Play as a Foundation for Hunter-Gatherer Social Existence," *American Journal of Play* (Spring 2009): 476–522.
4. Marjorie Shostak, *Nisa: The Life and Words of a !Kung Woman* (Routledge, 1990), p. 10.
5. Nicolas Guéguen, Sébastien Meineri, and Jacques Fischer-Lokou, "Men's Music Ability and Attractiveness to Women in a Real-Life Courtship Context," *Psychology of Music* 42, no. 4 (July 2014): 545–549.
6. Laurel J. Trainor and Erin E. Hannon, "Musical Development," chap. 11 of *The Psychology of Music,* 3rd ed., ed. Diana Deutsch (Academic Press, 2013).
7. Mary B. Schoen-Nazzaro, "Plato and Aristotle on the Ends of Music," *Laval Théologique et Philosophique* 34, no. 3 (1978): 261–273.
8. Trainor and Hannan, "Musical Development" (see reference 6), p. 425.
9. T. Nakata and S. Trehub, "Infants' Responsiveness to Maternal Speech and Singing," *Infant Behaviour and Development* 27 (2004): 455–464.
10. Tali Shenfield, Sandra Trehub, and Takayuki Nakata, "Maternal Singing Modulates Infant Arousal," *Psychology of Music* 31, no. 4 (2003): 365–375.
11. Sandra Trehub, Niusha Ghazban, and Mariève Corbell, "Musical Affect Regulation in Infancy," *Annals of the New York Academy of Sciences* 1337 (2015): 186–192.

12. Sandra Trehub, personal communication, April 22, 2015.
13. Raymond MacDonald, David J. Hargreaves, and Dorothy Miell, "Musical Identities," chap. 43 of *The Oxford Handbook of Music Psychology,* ed. Susan Hallam, Ian Cross, and Michael Thaut (Oxford University Press, 2009), pp. 431–440.
14. Marc J. M. Delsing, Tom F. M. Ter Bogt, Rutger C. M. E. Engels, and Wim H. J. Meeus, "Adolescents' Music Preferences and Personality Characteristics," *European Journal of Personality* 22 (2008): 109–130.
15. M. B. Holbrook and R. M. Schindler, "Age, Sex and Attitude towards the Past as Predictors of Consumers' Aesthetic Tastes for Cultural Products," *Journal of Marketing Research* 31 (1994): 412–422.
16. Adrian North and David Hargreaves, *The Social and Applied Psychology of Music* (Oxford University Press, 2008), p. 111.
17. R. Zatorre and V. Salimpoor, "From Perception to Pleasure: Music and Its Neural Substrates," *Proceedings of the National Academy of Sciences* 110, suppl. 2, June 18, 2013, 10430–437.

Fiddly Details

1. P. von Hippel and D. Huron, "Why Do Skips Precede Reversals? The Effect of Tessitura on Melodic Structure," *Music Perception* 18, no. 1 (2000): 59–85.

Index

About the Author

Dr. John Powell is a physicist and a classically trained musician, with naturally curly hair. He has given lectures at international laser conferences and played guitar in pubs in return for free beer. He prefers the latter activity. He holds a master's degree in music composition and a Ph.D. in physics, and has taught physics at the universities of Nottingham and Lulea (Sweden) and musical acoustics at Sheffield University. He lives in Nottingham, England.